Julia Davies & Guy Merchant

Web 2.0 for Schools

Learning and Social Participation

PETER LANG
New York • Washington, D.C./Baltimore • Bern
Frankfurt am Main • Berlin • Brussels • Vienna • Oxford

Library of Congress Cataloging-in-Publication Data

Davies, Julia.
Web 2.0 for schools: learning and social participation /
Julia Davies, Guy Merchant.
p. cm. — (New literacies and digital epistemologies; v. 33)
Includes bibliographical references and index.
1. Educational technology. 2. Web 2.0. 3. Online social networks.
I. Merchant, Guy. II. Title. III. Series.
LB1028.3.D28   371.33'44678–dc22   2009005538
ISBN 978-1-4331-0264-6 (hardcover)
ISBN 978-1-4331-0263-9 (paperback)
ISSN 1523-9543

Bibliographic information published by **Die Deutsche Bibliothek**.
Die Deutsche Bibliothek lists this publication in the "Deutsche
Nationalbibliografie"; detailed bibliographic data is available
on the Internet at http://dnb.ddb.de/.

Cover design by Joni Holst

The paper in this book meets the guidelines for permanence and durability
of the Committee on Production Guidelines for Book Longevity
of the Council of Library Resources.

© 2009 Peter Lang Publishing, Inc., New York
29 Broadway, 18th floor, New York, NY 10006
www.peterlang.com

Printed in the United States of America

CONTENTS

ACKNOWLEDGEMENTS

This book draws together work with new technology and literacy that has absorbed our attention and infiltrated our professional and private lives over a number of years. It seems appropriate to begin by thanking all those friends and colleagues who have contributed to our understanding of digital literacies and emerging technologies, particularly those who have helped us to conceptualise the learning that takes place in Web 2.0 spaces.

This specific work would not have been possible without the inspiration, enthusiasm and encouragement of Colin Lankshear and Michele Knobel, who have been generous and creative in all their comments on earlier drafts. We are also very grateful to all those who have let us write about their work and to use screenshots from their pages. These include Jackie Marsh, for permission to annotate her blog (www.digitalbeginnings.blogspot.com), and Jeroen Clemens and Anica Petkoska for allowing us to write about the MacNed Wiki Project. We would also like to thank the Perklets' Mum for talking with us (online) and for letting us share the Perklets with you. Serendipity (or was it Google?) led us to Mrs Cassidy and we are very grateful to her, and her class, for allowing us to use their blog.

Special thanks go to Rosa, for telling us all about the 'Fail Meme', and to Richard Giles (of virtuallylearning.co.uk) and Paul Rees (of Barnsley Local Authority), for allowing us to draw on material from the virtual world 'Barnsborough'. K-Zero kindly provided us with the graph that maps the metaverse, and Cathy Burnett allowed us to use material from her doctoral study. Finally we are extremely grateful to those teachers and pupils who have let us into their classrooms and allowed us to learn so much from them. Thank you.

PREFACE

Over the last five years, there has been a large-scale shift in popular engagement with new media. Virtual worlds and massive multiplayer online games attract increasing numbers, whilst online social networking and file sharing are commonplace. The changing nature of online engagement—often referred to as Web 2.0—privileges interaction over information. Web 2.0 applications promote widespread social interaction and have brought with them new notions of what it might mean to be literate in the twenty-first century. Indeed researchers have argued for a reconsideration of the definition of literacy, giving prominence and prestige to the so-called New Literacies (Lankshear and Knobel, 2006a), and an acknowledgement of the changing roles and responsibilities of literacy educators. Suggestions that literacy needs to be referred to in the plural in order to highlight multiple social applications, contexts and forms of texts have been aired in academic circles and beyond (Barton and Hamilton, 1998; Cope and Kalantzis, 2000; Street, 1993). Now a new-step change has occurred with the dramatic impact of technology and digital texts on our world. The increasingly widespread availability of broadband connectivity and the development of ever more responsive software have led to a greater recognition of the internet as a place for social interaction, a place for collaboration, and a place for strengthening and building social networks. Web 2.0 commentators have drawn attention to the 'social' and 'participatory' nature of contemporary life online (Lessig, 2004; Jenkins, 2006a;

Shirky, 2008), and innovators are just beginning to glimpse the educational possibilities of these new developments.

The term Web 2.0 was coined by O'Reilly (2005) as a way of referring to
a significant shift in the ways in which software applications were developing
and the ways in which users were adopting and adapting these applications.
New applications were tending to become more user-friendly and interactive
and provided greater scope for the development of online communities than
before. At the same time, it was becoming possible for those with relatively unso-
phisticated technical skills to create and share content over the internet. The
popularity of blogs as a medium for individuals and groups to publish and discuss
their concerns, news and interests (whether frivolous or serious) is testimony to
the popularity and everyday currency of the Web 2.0 phenomenon (Davies and
Merchant, 2007; Carrington, 2008, Mortensen and Walker, 2002).

O'Reilly used the term Web 2.0 to describe a phenomenon that was happening, rather than as a label for a specific piece of new technology. Thus the
term has meanings that are not entirely fixed—in fact, Berners-Lee (2006) has
disputed that there is such a thing as Web 2.0, since, he argues, the Web has
always been capable of the interactive features and uses associated with Web
2.0—'Web 1.0 was all about connecting people. It was an interactive space,' he
claims. Despite this ongoing debate, we find the term 'Web 2.0' useful, at least
because its employment emphasises the aspect of interactivity and a second wave
of enthusiasm for the internet that has met users' realisations of the potential of
Web software. The intense increase in the use of a whole range of newer applications that foreground interactivity and collaboration around shared content
merits the attention of educators. We need to evaluate it, consider current uses
and their educational implications as well as determine the role we could play
in developing Web 2.0 awareness and use. With so much free software available
and with so many young users already interested, now is a good time to invest
in all of this in education.

To date, discussion of the opportunities and indeed of the risks presented by
Web 2.0 development has been largely confined to the exploration of social and
recreational worlds. The purpose of this book is to open up discussion about the
relevance of Web 2.0 to educational practice. A central concern in what follows
will be to show how the new ways of communicating and collaborating that constitute digital literacy might combine with new insights into learning in ways that
transform how we conceive of education (Gee, 2004b). Pivotal to this exploration
will be the identification of new ways of thinking and being that are associated
with the new literacy and meaning-making practices that are emerging.

This is not a book that provides answers or solutions; our intention is simply to raise awareness and to pose what we think are important questions. Our thoughts are based on our own Web 2.0 research and exploration as well as on the work of the innovative teachers we have had the good fortune to work with. Some of their work is represented in the illustrative vignettes of practice that are included in the chapters that follow.

· 1 ·

EDUCATION AND WEB 2.0: TRANSFORMING LEARNING—AN INTRODUCTION

From the very beginnings of formal education, there has been an assumption that children and young people will attend a school or similar institution to learn and be taught there for a greater part of the day. Although ways of organising learners into teachable groups, the curriculum design that has guided what was to be taught, and the pedagogy employed have been subject to some variation, this basic concept of formal education has nevertheless served different societies and different social and cultural groups consistently well—if rather unevenly.

Learning in what have often been very diverse settings has normally been directed by teachers and orchestrated in various ways in determinations to ensure pupil engagement with new concepts and ideas as well as to support learning and collaboration with each other. Pedagogical strategies and resources, activities and tasks have been designed and allocated in ways judged to be appropriate to the age and level of different groups of learners. Technology, in terms of the sorts of tools that are familiar in these institutional settings, has been used to serve both systemic and instructional needs, to aid teachers in acts of demonstration, explanation and task setting. Tools such as blackboard and chalk, overhead projectors, electronic whiteboards and duplicated texts have been used to engage, occupy, regulate and sometimes oppress students by controlling their access to the stylus and slate, paper and pen, or latterly, the portable computer, and the scope of the sanctioned meanings of schooled literacy.

Information and Communications Technology (ICT) first became a curriculum priority in English-medium education systems in the 1980s. Since then, educators and others have repeatedly pointed out that the 'C' of ICT is often overlooked, and that the possibilities offered by rapid multimodal interaction between learners and teachers in different locations challenges the taken-for-granted fixities of traditional education (Bigum, 1998). The new technologies of the late twentieth and early twenty-first centuries have predominantly (but not exclusively) been used to replicate and reproduce older (pre-digital) practices. Despite their potential, new technologies are still regularly used to provide more polished performances of conventional practices—a phenomenon often referred to as 'old wine in new bottles' (Lankshear and Knobel, 2006a).

Learning bounded by the walls of the classroom, limited by interaction with those in the same location and regulated by opening and closing hours could be an outdated concept if we recognise the potential of new technologies. This is true from a number of points of view. Firstly, it is now widely recognised that a significant amount of learning takes place in out-of-school contexts (Carrington, 2008; Davies, 2004; Dowdall, 2007; Leander and McKim, 2003) and that informal learning is both important and worthwhile for children and young people. And secondly, as studies of classroom learning from Barnes (1992) to Mercer (2000) suggest, peer interaction can play a very important part in learning. If communication lies at the heart of the educational process, then new technologies present exciting possibilities. We believe that despite the commitment of many teachers to invest in the potential of new technologies, there is much work to be done in supporting others to become more confident and secure in their ability to provide creative, dynamic and meaningful activities that take the best of Web 2.0 applications to develop literacy teaching and learning in their classrooms. This requires proper and strategic investment from policy-makers and education managers who need to provide guidance, time and expertise for the use of technologies to develop current and new ways of working with ICTs.

What is Web 2.0?

As we outline in the Preface, the term Web 2.0 was coined by O'Reilly (2005) as a way of referring to a significant shift in the ways in which software applications were developing and the ways in which users were adopting and adapting these applications. So Web 2.0 does not refer to anything as specific as new hardware or a reconfiguration of the internet; it is a term that attempts to highlight a new wave and the increased volume of users who have developed new ways of using digital technology to interact with each other. These users' interests have been in online participation, in networking and in collaboration.

The emphasis has shifted to the 'Communication' of ICT capabilities partly because of developments in 'user-friendly' software. So, for example, one no longer needs expertise in html to publish and participate in web-based activity. From the point of view of the end-user, the commonly used phrase 'the read-write web' is useful in capturing the shifts in practice that O'Reilly (2005) outlines. Thus, as Berners-Lee argues (2006), although the Web has always had the *potential* capacity to enable intense person-to-person, or one-to-many interactivity, *more people* are now able to participate—and *more types* of people can be involved—those who know about many things other than technology. We have, therefore, seen a huge increase in sites where users do not 'see' or acknowledge the technology they are using but simply 'get on' and communicate. All this further reflects a shift in emphasis in which web-based activity is no longer simply about storing and accessing information—an enormous Borgesian library—but more about interaction, a place in which individuals 'converse', reacting to each others' ideas and information and adding to the stock of knowledge. If Web 1.0 could be conceived of as an enormous encyclopaedia, containing information to locate and consult (authored by specific, known and carefully selected experts), then Web 2.0 is exemplified by Wikipedia, a growing repository of user-generated material, dependent on the collaborative endeavour of shared expertise, contribution and regular updating (by many self-selected and anonymous authors). And this is the essentially participatory nature of what some see as a new era in internet use.

Lying behind the 'read-write web' are important developments in software design and ways of using computing power that assist and enhance the processes of social networking. An important aspect of these developments relates to the way in which Web 2.0 applications can learn *about* and *from* their users, their capacity feeds this back to those users in mutually beneficial ways. O'Reilly (2005) gives the example of the Google search engine, which, by ranking searches by the number of hits, feeds back the information it collects from ongoing search activity. A more sophisticated example of this is the Amazon online bookstore, which not only remembers us when we return and archives what we have previously bought but, by matching our buying habits with others, is also able to suggest what else we might be interested in reading. These examples are offered simply to illustrate the Web 2.0 phenomenon although in themselves they are not sufficient to develop any sophisticated claims. Many readers will be aware of the shortcomings and loopholes in the Google ranking system, as well as the argument that Amazon isn't just being helpful in recommending new books—it is, after all, a business enterprise that thrives on selling merchandise. Indeed this is a timely reminder that ethical and commercial issues are never far away when we look at web-based environments.

In summary then, the distinctiveness of Web 2.0, of which social networking software (SNS) is a major part, can be attributed to a shift of emphasis towards user-generated content coupled with mechanisms that enable and enhance user interaction. User-generated content can thematically vary enormously, and users may consist of individuals or groups of people. Thus one can find blogs devoted to recipes for cupcakes, or about the latest shoe fashions, or they may be educational sites that exchange ideas for teaching, or even shared photostreams that exhibit portrait photography. Web 2.0 spaces can exploit the affordances of different media from text, to still image, to moving image, to sound, and to any combination of these. At the same time, user interaction can be encouraged by applications that allow for such features as profile pages, messaging facilities, group formation and category tagging, all of which help us use and sort through the content in different ways, prioritising some aspects of the content over others. More sophisticated sites may allow users to see which of their friends are online, provide information on the latest changes to favourite sites and allow the choice of modifying or personalising their home pages, changing their look and the features included. Many sites, conversely, allow us to obscure our offline identity in all our communications, either by using avatars or by simply letting us contribute content without revealing specific authorship to everyone (such as on some wikis).

From this brief overview, it should be clear that Web 2.0 presupposes a more active user who is encouraged to design an online presence (an identity, or even multiple identities) and to participate, to a greater or lesser extent, in a community of like-minded users. Whether or not the social networks produced can be described as 'communities of practice' (Wenger, 1998) or 'affinity spaces' (Gee, 2004b) and how we can describe the informal learning that takes place in Web 2.0 environments are areas we explore in this book. We are interested in considering and helping readers to think through the implications of Web 2.0 not just in terms of redefining literacy and the literacy curriculum, but also in terms of broader spheres of education, its impact on our everyday lives and on how we see our world and our place within it.

Some key characteristics of Web 2.0

As we have seen, Web 2.0 spaces have a number of characteristics. O'Reilly has a lengthy list and others (e.g., Cagle, 2006) have developed similar lists. We recognise that since Web 2.0 is best described as a developing trend or attitude (Lankshear and Knobel, 2006a), it is likely that some but not all of these features may be present in a single Web 2.0 space. However, in order to capture the essence of Web 2.0, we find it useful to refer to four characteristic features, and these are used at various points in this book. They are listed and explained below but will become clearer as the book progresses.

1. *Presence*—Web 2.0 spaces encourage users to develop an active presence through an online identity, profile or avatar. This presence is recognisable by others but may develop over time. Active presence is recognised by updating, interacting and in some cases alerting to show that a user is online. Many users develop a sense of self across a number of spaces—such as through one or more blogs, in a Flickr stream, in eBay and on YouTube (Davies, 2008).

2. *Modification*—Web 2.0 spaces usually allow a degree of personalisation such as through the design of the user's home page and personal links, or through the creation of an on-screen avatar. Web 2.0 spaces may also be 'mashable', or interoperable. The API (application programme interface), which acts as a sort of handshake between programmes, allows users to link one application to another or import objects and features from one space to another—such as embedding images from Flickr in a wiki, or a YouTube video in a blog and so on. This kind of transportability allows the development of an online presence across spaces as referred to in the previous point.

3. *User-generated content*—Web 2.0 spaces are based upon content that is generated within and by the community of users rather than provided for the community by the site itself. That is to say, YouTube, for example, provides a template and plenty of online space for its users, but it is the users who supply the videos, comments and other content. This, of course, does not mean that participation is not possible if users do not generate content. For example, there are many users of YouTube who do not upload videos or comment on the site—but they may embed video codes from the site onto their blogs, or they may cite URLs in their MySpace, for example. In this way Web 2.0 users are producers as well as consumers.

4. *Social participation*—Web 2.0 spaces provide an invitation to participate. This derives, in part, from the above three points. Rating, ranking and commenting are all ways of giving and receiving feedback and developing content, whereas

features such as friend lists, blogrolls and favourites become public displays of allegiance (Donath and boyd, 2004). Just as user-generated content makes us both producers and consumers, so with social participation we are simultaneously both performers and audience.

This list of characteristic features is not intended to be exhaustive and should not be seen as a set of criteria to judge whether a site is Web 2.0 or not. However, we have developed a sense of such spaces and note the above as tendencies, as features that seem to characterise Web 2.0 for us. The list has proved useful in identifying some of the key ways in which the various Web 2.0 spaces that we examine involve their users or members and how they promote a sense of community and interaction.

Why should we be interested in Web 2.0?

It appears that social networking sites and other Web 2.0 spaces have captured the public imagination. Media reports about the changing favours of MySpace and Facebook, the earnest discussion about the rise of citizen journalism through newsblogs, the scepticism and adoration that Wikipedia seems to receive in equal measure, all underline the fact that the Web 2.0 phenomenon is both popular and contentious. Popularity can be gauged by the headline statistical information that we are presented with on an almost daily basis (something that we discuss in Chapter 3), as well as by the everyday awareness and use of these new web-spaces. But these new online spaces are not without their critics. Familiar moral panics about 'alienated techno-subjects' (a term coined by Luke and Luke, 2001) abound. Cyberbullying, misinformation, copyright infringement and a whole host of other worries fuel the risk discourse that surrounds Web 2.0, and we discuss these in Chapter 9. Whatever the significance we attach to these risks, their high profile suggests that we ought to be addressing them in educational contexts. However, we also want to argue that there are other, equally significant reasons for exploring the relevance of Web 2.0 for schools and other educational institutions. Some of these are set out below:

- Many children and young people are already engaged in Web 2.0 practices;
- Important kinds of learning can be developed in Web 2.0 environments through knowledge sharing and distributed cognition;
- Web 2.0 users are developing new online social practices that are likely to become more useful in the work and leisure settings of the future;

- Web 2.0 and online social networking practices can be enjoyable—they motivate the young and can also be attractive to teachers;
- New webspaces are dependent on new literacy practices, providing real-world environments to practice and develop these;
- The risks and opportunities of life online can be carefully explored in educational environments;
- Web 2.0 environments give voice to participants and suggest new possibilities for social engagement and citizenship;
- Collaboration and criticality can be developed in Web 2.0 engagement.

In summary, we would like to argue for the significance of Web 2.0 as a way of motivating teachers and students, as an arena for developing new knowledge and skills (see Bryant, 2007) and as a medium that promotes new kinds of collaborative learning. However, it is, at the same time, important to acknowledge that the reasons we have set out as a rationale for Web 2.0 work are not unassailable and are themselves based on certain assumptions. Figure 1 presents our claims alongside some counter-claims. What is striking, here, as we shall see, is that with close examination the counter-claims themselves also provide their own justifications for an educational focus on Web 2.0.

The chart shows, for example, that although many claim that we do not need to address Web 2.0 in schools because learners spend so much of their time already involved in Web 2.0 and are thus more confident and skilled than their teachers—we could counter this argument by explaining that, in fact, there is a digital divide. Not all young people are able to engage in Web 2.0 practices, they often run into problems when using it or use it in undeveloped and repetitive ways. Teachers have a role to play in guiding their students' use of technology. We further argue that Web 2.0 can build on students' current funds of knowledge and appreciatively engage learners who are often disenchanted with traditional school approaches to literacy teaching.

In what follows, we evaluate the characteristic features and educational potential of some significant Web 2.0 practices. The next chapter, Chapter 2, provides the theoretical underpinning for this work; the subsequent chapters go on to explore some of the more popular ways in which new online spaces are being used. In doing this, we have been obliged to make some hard choices about what to include and what to omit. In the first instance, we have been guided by the practices that we are most familiar with, either through our direct participation or through our research. Sometimes this means that we provide an overview of a

CLAIM	COUNTER-CLAIM
Large numbers of school-age children are already engaged in Web 2.0 practices	A significant number of school-age children do not have access to these practices (the digital divide)
Web 2.0 promotes knowledge sharing and distributed cognition	Most young people's online practices are repetitive and limited in ambition
Young Web 2.0 users are developing new online practices that will be transferable and useful in the future	Web 2.0 practices may be trivial and unproductive
Online social networking and related Web 2.0 practices are enjoyable and motivating for young learners and their teachers	Children, young people and their teachers may have had negative experiences of new technology and find it 'difficult'
New webspaces involve new literacy practices and provide real and meaningful experiences	Literacy is an important asset that is being gradually eroded by new media
The risks and opportunities of life online can be explored through engaging with Web 2.0	The internet is a potentially dangerous place—children and young people need to be protected
Web 2.0 spaces give voice to participants and suggest new possibilities for social engagement and citizenship	Web 2.0 spaces create an illusory sense of social engagement and actually create passive citizens who can easily be surveilled
Collaboration and criticality can be developed with Web 2.0 engagement	Web 2.0 creates isolated techno-subjects who use the internet for vanity publishing and recycle old ideas

Figure 1. The educational rationale for Web 2.0: claims and counter-claims.

topic area, as in Chapter 3, which addresses the practice of blogging and does not discuss a particular piece of blogging software. On other occasions, we have chosen to focus on a specific site, using it as an example of that general area of practice. So, Chapter 4 concentrates on Flickr, one of the more popular photo-sharing sites. Providing an in-depth view of the way in which Flickr works is, we feel, more productive than attempting a comparative overview of photo-sharing sites.

In a similar way, Chapter 5 provides a critical examination of video sharing by focusing on YouTube. We feel that this focus on YouTube raises important issues for educators, particularly in view of the fact that this service is quite commonly blocked in educational establishments. We nevertheless argue that YouTube has a great deal to offer educators and that there are important reasons to argue for a relaxation of the embargoes against it in most schools and colleges (within the UK, the USA and beyond). Chapter 6 looks at the whole area of music sharing, in the belief that this quite distinct area faces so many significant issues concerning copyright and legality that it warrants a rather different kind of treatment.

Whether or not one classifies virtual worlds as Web 2.0 is a contentious issue. Given the potential of virtual worlds for promoting social participation and distributed learning, we decided to include this topic. We recognise that although there is a lot to be said about virtual worlds, research into their educational use is still in its infancy. As a result, we have narrowed the focus in Chapter 7. We do not address the topic of informal and formal learning in Second Life (probably the best-known and most popular virtual world), choosing instead to concentrate on virtual worlds for children—in particular the issues raised through a case study of virtual gameplay in a school context.

In Chapter 8, we turn our attention to wikis, taking a general view of how wiki technology provides opportunities for collaborative text production in educational contexts. We illustrate this with reference to a number of small-scale initiatives that tease out some different dimensions to collaborative authoring and wiki production. Chapter 9 provides a synthesis of the ideas about learning developed throughout the book and charts the way forward for Web 2.0 work in schools. The Conclusion returns to some fundamental questions about Web 2.0 and examines some of the underlying assumptions about new technologies and schooling.

New technology is a difficult area to write about, not least because what is new at the time of writing can often become commonplace or forgotten by the time of publication. Web 2.0 is particularly challenging because of the rapidity of its change and the nature of the industry. By restricting ourselves to writing about what we are familiar with we aim to give the topic a fair treatment. We also think that the particular software we refer to has been around long enough and is popular enough to suggest that it is unlikely to fade overnight. However, we are well aware that time and duration are very different in new virtual environments!

In this book, we do not present Web 2.0 as a panacea for all evils; we are aware of the barriers to change and of the need to exercise caution and control when working with Web 2.0. As with any resource—books, magazines or the trip to out-of-school places—teachers need to feel confident they have made the right choices and are able to support students in their learning. We offer information about a range of Web 2.0 sites and possibilities for teaching, discussing both the advantages and disadvantages of these for teachers and their learners. Finally, however, we are writing with the awareness that whatever happens, young people across the world are engaged in and will continue to be engaged in Web 2.0 practices. They can benefit from the guidance of their teachers, parents and other mentors—and we hope that this book will provide some of the confidence, information and enthusiasm needed to begin investigating new possibilities offered for teaching and learning in the new Digital Age.

WEB 2.0 AS SOCIAL PRACTICE

In this chapter, we explore theoretical perspectives that are useful in developing an understanding of Web 2.0 practices and their educational potential. This offers the reader an introduction to ideas that are addressed in more depth later in the book by looking at key examples of Web 2.0 work. In this sense, what follows provides an important underpinning to subsequent chapters. We begin by considering Web 2.0 activity as textually mediated social practice, using this idea to explore the multimodal nature of the dynamic screen-based texts that are created. This leads to a consideration of the literacies used in interaction and collaboration that in turn promote social participation and the creation of social networks. The chapter concludes with a consideration of the identity work that are brought to the fore in Web 2.0 spaces.

Web 2.0 services provide a context for social practices that are based upon people's contribution to, and joint construction of, web-based texts. They could, therefore, be characterised as a set of cultural and transcultural communicative practices—practices that are dependent upon shared and negotiated models of social interaction. In this analysis, we draw on the work of the New Literacy Studies that describes textual production and consumption in terms of the social practices involved. These are social practices as defined by Street (1993) and Barton and Hamilton (1998) that relate to the norms, values and so on that inform processes of interaction. From this point of view, the term 'practice' is used to capture the sociocultural and political forces that pattern interactions, rather than the specific interactions themselves. In the theoretical model developed by Barton (1994), the term 'events' is used to describe the specific activities that take place around texts (see Vignette 1, for some examples). These events can be described in terms of the kinds of participation that are involved. In some cases, participation may be fairly limited (as in the case of skim-reading a friend's recent blog posting); in others, participation may be fully collaborative (when jointly authoring a wiki or engaging in synchronous chat). Whatever the level of participation, the text plays a central role in mediating the interactions that take place, making them what Barton describes as 'literacy events'.

Reflecting on what has been learnt from research guided by New Literacy Studies, Barton (2001) concludes:

> Nearly all everyday activities in the contemporary world are mediated by literacy and that people act within a textually mediated social world. (Barton, 2001:100)

This is a powerful observation, one that warrants some careful attention in relation to Web 2.0. If Web 2.0 can be seen as an extension or even as a central aspect of some people's everyday activity, then we can clearly describe online interaction as being textually mediated—and indeed this is the direction we will be taking. But first, since Barton's summary highlights the role played by literacy, this deserves further discussion.

Text, literacy and social practice

The observations of daily life on a commuter train in Vignette 1 illustrate Barton's point about a textually mediated world and provide examples of everyday literacy practices that we will draw on in the rest of this chapter.

> *The commuter train is full—there is no option but to stand. To either side of me there are facing seats around shared tables. On one side a young man is sorting through old train tickets, talking across the table to his friend opposite. Next to him, someone has started work already, entering figures into a spreadsheet on a laptop. The conversation that the young man is engaged in draws to a close as his friend pulls a book from his bag and a pencil to mark the text. The book is* The Discourse of Design. *Next to the reader is a young woman. She is texting her friends. Looking down, I notice that she is commenting about how the train is*

'rammed' and how fortunate she is to have a seat. At the table across the way, there is a young woman in the corner seat with a list and a stack of Christmas cards. She signs and seals them one by one. Across the table from the Christmas cards is an older man. He reads a letter (it's about a property transaction). He turns the letter over and begins to draft his reply on the reverse side. Next to him is a female student reading a familiar-looking assignment brief. And opposite her, next to the young woman with the Christmas cards, someone is doing a Sudoku puzzle from the morning newspaper.

Vignette 1—Literacy: The textually mediated social world

Although by no means all of the activities described involve digital texts and we are aware that the general processes we illustrate here may well be common, beginning with the idea of textual production and consumption, we can see how this is reflected in many of the events described above. Whether the activity involves lettered representation (the Christmas cards), or numerical representation (the Sudoku puzzle), or conversation, the participants described are each involved in the co-construction of meaning, alternating between active and passive roles. Even in the case of book reading, the participants are not simply passive consumers; in the vignette, this is illustrated materially by the reader, with pencil in hand, ready to mark significant parts in the text—but, of course, on another level, the reader is constructing or producing meaning as he reads.

The perspective taken in this book is that online activity can usefully be described as a set of textually mediated social practices. Web 2.0 spaces can be analysed in terms of textual production and consumption as well as the relationship and interplay between the two. Photo sharing, i-m and blogging, for example, all involve participation through engagement with new media texts. Media, in this context, refers to the internet (hence the use of the term 'new'), and the pages, sites or platforms that provide the immediate context for interaction can be seen as texts (Buckingham, 2003). These texts are *dynamic* and *loose-bound* in character. Dynamic because it is often difficult to define a point at which they are complete, and loose-bound because the boundaries that demarcate one text from another are fuzzy. For example, a blog is, by definition, often updated and in this sense is always 'work in progress'. Similarly, each post may seem fairly complete, but the successive posts can be related. Whether the unit of analysis for such a text is a whole screen, a specific post or the whole site, including archives, is a contentious issue. And then, when we turn to a consideration of comments added over a period of time to individual posts and the various hyperlinks that may be employed, we begin to get a sense of the ways in which these texts are loose-bound.

As we saw in Vignette 1, literacy events are always situated in broader contexts, and they often involve other semiotic modes. So, the text message gathers its meaning from the knowledge that the sender of the message is on a journey, what it's like when a train is crowded and the specific idiomatic nuances of the expression 'rammed'. In a similar way, the examination of the train tickets is located in the conversation that surrounds them and refers to them. Work online operates in a similar way, often referring to other sites, events in the real world, previous conversations or face-to-face interactions. Moreover, text is often much more than communication through the written word, since meanings are constructed also through sound, image and gesture as well as their interaction.

Another key feature of online texts and, therefore, of those that can be categorised as Web 2.0 is their multimodal nature (Burn and Parker, 2003). The multimodal perspective, developed by Kress and Van Leeuwen (see, for example, Kress and Van Leeuwen, 1996; 2000) is particularly useful in describing and understanding the subtle interplay between different expressive modes such as sound, image and written word that characterise screen-based text (see Figure 2). A theme explored in this book is the specific role played by literacy in the social interactions that are a central theme in Web 2.0 networks. We take the view that an understanding of literacy is key to unlocking the educational potential of Web 2.0. However, 'literacy' in discussion and debate about online texts has become a contentious term, and it is this that we turn our attention to next.

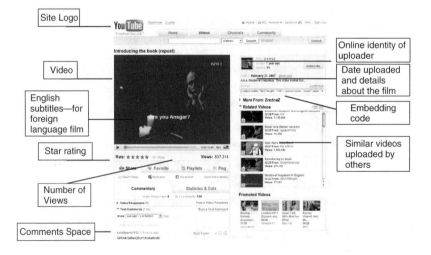

Figure 2. YouTube page.

In *Literacy in the New Media Age,* Kress (2003) argues that 'lettered representation' is a central defining feature of literacy. He goes on to develop a view of new media that illustrates how we construct meaning through a variety of modalities that complement, augment or replace written representation in different communicative contexts. This view of literacy and its place in reading a complex screen such as the YouTube page, shown in Figure 2, is not without its limitations but is central to an informed understanding of digital literacy. The page shown in Figure 2, as can be seen, is highly complex and requires skills additional to those taught for paper-based reading of non-html written text. We have, quite evidently, labelled only a minority of features, just to give an idea of what different regions of the page refer to. One needs to know which text is hyperlinked and which text is essential to video viewing. Kress's definition of literacy is limited in the sense that it seems to distinguish between the use of letters and other symbols such as icons, pictograms and numbers as well as visual features that indicate corporate branding, navigational features and hyperlinks. However, we argue that these fall within the domain of written or symbolic representation and are not concerned solely with letters and the related skills of decoding. As Harris (2000) suggests, new technology begins to blur the 'conventional boundary lines which treat iconic symbols as non-words'. We suggest that, despite these concerns, it is important to place written (symbolic) representation at the heart of any definition of digital literacy. Later we will illustrate how, even in Web 2.0 environments that are organised around sharing videos or music, interaction is still largely driven through the written word.

Participation in social networks

By placing the emphasis on literacy as a social practice, attention is turned from a consideration of how individuals read or construct meaning in isolation to an understanding of the ways in which meaning is negotiated through interaction and in context. The individuals and the textual practices of the fellow travellers on the commuter train described in Vignette 1 above are all located in social networks, sometimes present and sometimes remote. There will no doubt be recipients of the Christmas cards, an audience for the spreadsheet, an assignment to be discussed, submitted and marked and so on. These and other literacy practices play an important role in the social networks that individuals participate in. Literacy practices are, of course, sometimes central and sometimes quite incidental to social networks, and this is equally true of online and offline contexts.

A number of Web 2.0 applications are categorised as social networking software (SNS), thus highlighting one salient feature—that of establishing and maintaining connections between participants. The most obvious example of this is Facebook, whose owner has described his project in terms of mapping 'the set of connections that everyone has in real life' (Zuckerberg, 2007). In this sense, Zuckerberg is taking up and further developing the early success of FriendsReunited in harnessing computational power to enhance contact and communication between people. A number of authors have used models of social networking theory to comment on and map the sorts of connections now being made between people in online interaction and through online communities, and to chart and theorise the sort of changes that may be occurring. One of the most influential of these is Wellman (Wellman, 2002; Wellman and Hogan, 2004; and Wellman et al., 2005).

Wellman and his colleagues, through a number of empirical studies, have developed a theory of social networks that shows how trends in the use of new technology are embedded in the wider and changing landscape of social interaction. Wellman et al. (2005) use a tripartite model to chart the shift from traditional close-knit communities to glocalised networks (which are both global and local) and on to what they call 'networked individualism'. Close-knit communities (or groups) are collocated and most communication is face-to-face, whereas glocalised networks are based on place-to-place communication. Wellman goes on to suggest that, in the case of networked individuals, the person rather than the place becomes the locus of connectivity. He uses the mobile phone to exemplify this.

> With the internet and the mobile phone, messages come to people, not the other way around. Individuals are connected by their phones, but their phone is not tied to a place and its environment (such as family or office). (Wellman et al., 2005:1)

This is illustrated by the text-message exchanges described above in Vignette 1, since although the messages being sent make specific comment on the immediate location, the destination of these messages is difficult to guess. The primary focus is on the connections between the participants, rather than their specific locations.

In an earlier article, Wellman (2002) suggested that, although his tripartite theoretical model signals directional changes in patterns of social organisation, it is quite possible that many people inhabit a social world that is actually a *mixture* of groups and networks. However, Wellman also observes that the move to networked individualism raises important questions about the emergence of new social identities. He poses two questions that are pertinent to our discussion here:

- Will networked individualism deconstruct holistic individual identities? (Wellman, 2002:16)
- To what extent does the Internet reduce the importance of traditional social organising criteria such as gender, social class, ethnicity, language, life-cycle stage, and physical location? (Wellman, 2002:16)

We return to these questions towards the end of this chapter when we turn our attention to an exploration of identity; but first it is important to take our discussion of social participation a little further and to discuss how it relates to learning.

Social participation and learning

In the previous section, we argued that meaning is negotiated through interaction, and that this holds true for the production and consumption of offline and online texts alike. We have deliberately chosen the term 'social participation' as a way of describing how this happens because it fits well with current thinking in media studies (Jenkins, 2006a) as well as with theories of situated learning (see Murphy and Hall, 2008, for an overview). In this section, we highlight some of the key themes in a participatory account of learning.

In stark contrast to the idea that contemporary life is fragmented and is typified by the loss of a sense of value, purpose or community (Bauman, 2003) and to the associated discourses about youth as isolated and disaffected techno-subjects, the work of authors such as Jenkins (2006b), Levy (1997), Rheingold (2002) and Shirky (2008) has championed the notion that new technologies can create increased levels of participation. It has been argued that rather than rendering new citizens passive, technology provides opportunities for new kinds of social organisation and even resistance—for instance, through culture-jamming and through the mobilisation of smart mobs (Rheingold, 2002; Shirky, 2008; Lankshear and Knobel, 2006a). In his writing on this topic, Jenkins regularly uses the term 'participatory culture' to capture this theme and to suggest the importance of audience participation and collective intelligence in contemporary mediascapes.

In a paper written for the MacArthur Foundation, Jenkins (2006c) defines a participatory culture as one:

1. With relatively low barriers to artistic and civic engagement
2. With strong support for creating and sharing one's creations with others
3. With some type of informal mentorship whereby what is known by the most experienced is passed along to novices
4. Where members believe their contribution matters
5. Where members feel some degree of social connection with one another (at least they care what other people think about what they have created). (Jenkins, 2006c:7)

Participation is also a key theme in contemporary theories of learning. Sociocultural accounts tend to emphasise the importance of social interaction either from the point of view of what individuals learn *from* joint endeavour (what Rogoff [1995] refers to as 'participatory appropriation') or from the point of view of what is learnt *in* joint endeavour (participation in a 'community of practice' [Wenger, 1998]). Participation, for Wenger, does not simply equate with collaboration, and this is partly because it includes conflict as well as the dynamics of power relations; more importantly, however, it involves a two-way traffic between the individual and the community, or as he puts it: 'participation in social communities shapes our experience, and it also shapes those communities' (Wenger, 1998:56). Although influenced by these ideas, this book does not apply the concepts of 'communities of practices' or Gee's 'affinity spaces' (Gee, 2004b) to Web 2.0, preferring instead to see these sites as 'spaces of reflection' (following Wegerif, 2005), or as places in which multiple voices and competing perspectives coexist. For us, then, social participation in Web 2.0 has the following features:

- it involves online communicative interaction in a shared space related to a joint endeavour or social object;
- although it may involve collaboration, it is not dependent upon it—there maybe conflict or opposition;
- it has distinctiveness because the interaction is predominantly online—it, therefore, does not depend on co-presence (in time and place), and it does not depend upon a face-to-face relationship, although neither are precluded;
- it takes place in an online environment that is shared, and in which knowledge and learning are distributed;

- levels of engagement are dictated by individuals who shape and are shaped by the community in different ways.

Descriptions and theories of learning through participation tend to emphasise how new learning involves taking on new identities. For example, Wenger (1998) places a strong emphasis on identity, whereas Gee (2004a; 2004b) provides a model for identity change. Seminal studies of online behaviour, although not specifically concerned with learning, have made an important contribution to our understanding and it is these that we turn to next.

Identity and social being

Participation in meaning-making practices involves taking up particular identities. In the context of Vignette 1, which we used to illustrate the social practices of literacies, it would be relatively easy to suggest the ways in which the individuals involved engaged in activities that either positioned them or invited them to adopt particular identities (co-worker, partner, student and so on). In a similar way, the study of online communication repeatedly shows the extent to which interaction is interwoven with identity performance (Turkle, 1995; Markham, 1999; Sunden, 2003; Thomas, 2004). There are a variety of accounts for this phenomenon. The most persuasive revolve around two positions. The first is related to the affordances of the technology itself, and the second, following Wellman, is based on an account of the wider social conditions in which new technology is embedded.

The first explanation derives from an acknowledgement of the impact of the recent and rapid increase in alternatives to face-to-face communication. Although the facility to communicate with those who are not co-present, or who are geographically remote, is arguably as old as writing systems themselves, new technology undoubtedly provides us with a range of tools that enable us to interact in different ways within more diverse and dispersed networks than previously imaginable. Although we are beginning to see some developments in the use of webcams and internet-based video-conferencing, most online interaction takes place though digital writing—a relatively lean medium that is stripped of the prosodic features of oral communication and the paralinguistic features of face-to-face communication. As a result, it seems important to signal who we are and what we feel in alternative ways. From this perspective, new tools for communication provide a context for new kinds of identity performance or, as some commentators argue, have helped to create a new kind of person (Thomas, 2004).

A second explanation for this growing interest in identity and new media derives from a broader view of the contemporary social context. Wide-reaching changes in the economic, political and social order, which have had both global and local impact, have produced both the necessity and the desire to create and maintain new kinds of identity (Giddens, 1991). The rise of a new capitalism (Gee, 2004b) with a global reach has given rise to a system in which it is less likely that goods are produced and consumed locally, and more likely that production is coordinated across locations and that goods are marketed to consumer types rather than geographical locations. This sort of arrangement requires the development of particular communicative tools but more pertinently leads to the emergence of new social identities—identities that are more accurately defined by lifestyle, media consumption and affinity spaces than by the more traditional markers of race, class, gender and place (Kress and van Leeuwen, 1996).

Whichever explanation we find the most persuasive, it seems an inescapable fact that the landscape of communication has changed and indeed continues to change, and that fact in turn raises new possibilities for constructing and performing social identity. In Merchant's study of teenagers in chatrooms (Merchant, 2001), participants established allegiances around favourite TV shows, humorous websites and popular music. Later, in a study that focused on the development of primary school children's narrative writing through email exchange, he noted how the role-play characters they constructed hybridised their interests in fashion and relationships with their media interest in the fantasy genre (Merchant, 2004). This work led to the development of a model that accounts for the ways in which children and young people perform both *transient* identities (e.g., those associated with popular culture, iconic figures and objects or fandom) and *anchored* identities (e.g., those which reflect gender, position in family, religion, age, social class and geographical location). Of course, these two different kinds of identity performance interrelate and overlap. So, for example, the way in which gender identity can be signalled in online interaction by making reference to musical preference, sporting icons and other interests becomes a particularly fruitful area for further exploration. This work suggests that what Wellman refers to as 'traditional social organising criteria' (2002:16) are still significant in the individual's sense of themselves as a social being.

Similarly Davies' work (2004) looked at ways in which tweenies experimented with a range of identities when playing online with so-called Babyz; they emailed their babyz to each other and set up schools and hospitals for them in their own personal websites. They managed joint sites and in so doing took on a range of roles using different types of 'voice' to perform the different tasks involved in being 'mother', 'nurse', 'website manager' and so on. Her later work looking at teenagers acting as 'witches' online also explored ways in which the young can take on new kinds of discourses that move in and out of different types of constructed online identities (Davies, 2006). Using this kind of example, Davies has (2009) demonstrated just some of the ways in which the internet can empower individuals to disseminate their voices—often blending the playful and earnest—and the young have found this particularly seductive, being quick to seize opportunities to explore the boundaries of possibility for the taking on of different kinds of 'transient' identities.

Davies and Merchant's (2007) collaborative study of academic blogging adds to the theme of identity performance through an exploration of how this manifests itself in self-publication and through online social networking. This auto-ethnographic study illustrated how academics textualise themselves in online communication, performing carefully managed practices of identity using the affordances of social networking software. This work seems to suggest that what Wellman referred to as 'holistic individual identities' (2002:16) may be a cultural construction that is no longer relevant as those with an online presence may perform multiple identities—for instance, as an academic on a blog, a photographer on a photo-sharing site, an old school friend on Facebook and an avatar in a virtual world.

Performing identity—networks and cultural artefacts

Our brief exploration of anchored and transient identities in the previous section began to point to the role played by cultural artefacts in identity performance. We gave the example of iconic figures and objects as examples of this, but it is also the case that what we consume, the spaces we occupy and the ideas we hold act as touchstones for our identities. SNS theorists (e.g., Kinsella et al., 2007) sometimes refer to 'social objects' and identify how many Web 2.0 applications involve networking around a social object (such as photographs and video clips in Flickr or video clips in YouTube). We prefer the broader category label of 'cultural artefact', suggesting as it does, ideas, objects and activities that are in one way or another produced and reproduced and are perceived as holding meaning within a specific culture. This allows for a broader and more dynamic definition than the term 'social object' evokes.

In online practices, identity is performed in social networks. Identity and interaction are always textually mediated—and often, as we have argued, through lettered representation. Cultural artefacts are important signposts in these networks, whether they are the books one is currently reading (displayed on one's website), the music you are listening to or the websites you have visited. In the next chapter, we will show how these are brought together in the world of blogging.

· 3 ·

OUT THERE: GOING PUBLIC WITH BLOGGING

As we saw in the first chapter, the weblog or blog is one of the most well-established and well-known Web 2.0 applications. A blog is simply a website that an owner or 'blogger' can update on a regular basis. Updates are normally date-posted and displayed in chronological order with the most recent entry or posting shown at the top of the page. Since regular posting is a key feature of blogging, most blogging software is quick and easy to use. A blogger can, therefore, publish news, views, notices of events, thoughts or random observations on the internet within minutes. The enormous variety of blogs—often collectively referred to as the blogosphere—includes serious and trivial material and is a largely unregulated clamour of individual and group voices. Depending on one's point of view, this can be seen as a fascinating diversity of human expression or as a confusion of unfiltered information and opinion.

Although some early commentators compared the blog to journal or diary writing, the only real similarity is the idea of regular dated entries. Admittedly some bloggers use their blogs in this kind of way, but a significantly larger group takes advantage of the affordances of the blog for different purposes. So, for example, the BBC news blogs provide fairly reliable updates on news items;

Henry Jenkins Fan Fiction reports on the U.S. academics' thoughts and publications on popular culture and new media; whereas the IRA blog provides digests and news on the developments in the field of literacy. These and many other blogs sit alongside the published ideas, views and interests of less well-known individuals and groups.

What is a blog?

A 'weblog' or 'blog' is one of the most well-established and well-known kinds of social software. A blog is simply a website that an owner or blogger can update on a regular basis. The word blog is derived from 'weblog'—quite literally a log or record of information presented as a date-ordered template, or as Walker describes it: 'A frequently updated website consisting of dated entries in reverse chronological order' (Walker, 2003). These entries, referred to as posts, are usually titled, with the most recent posting at the top of the screen. As mentioned above, since regular posting is a key feature of blogging, most blogging software is designed with ready-made templates and user-friendly interfaces that are quick and easy to use. A blogger can, therefore, publish on the internet within minutes.

Just as there are many different types of novel or types of poetry, so too there are many different types of blog. Every blogger will have his or her own reason for publishing, and there are probably more differences across blogs than there are similarities. Interestingly, it is not so much semantic content that characterises blogs, but rather their textual layout. That is, they are formatted, by default, in a reverse chronological order and typically contain side-bars with links, and a clear title across the top. Although blogs have some of the organisational features of a diary, despite their date-formatted structure, it would be inaccurate and over-simplistic to describe them (as some have done) as an 'online journal'.

We have looked at a great many different types of blogs and below identify just some of the many different uses and types we have looked at:

TYPE	EXAMPLES
Academic updates	Many academics keep blogs that allow them to share and explore ideas in a public arena. The use of a blog allows them to publish quickly—sometimes getting feedback and help with ideas. Academics have been quick to see the potential of blogs for networking and the opportunity to write in a way that is a hybrid between private research notes and more polished publication.
Community art projects	One of the most renowned art project type blogs is 'Post Secret' where many people are invited to send postcards with a secret written on the back, to a blogger who photographed and uploaded one 'secret' daily. It is now quite common for artists to blog the development of projects (see, for example, http://www.beckybowley.blogspot.com/).
Citizen journalism	This is where 'ordinary' members of the public, rather than professional journalists, keep blogs to report or comment on current events. Typically they offer alternative viewpoints to the official news channels (such as TV news and newspapers). Sometimes they focus on the purely local and thus provide a more locally relevant news service. A key feature is often the reporting via mobile technology—such as mobile (or cell) phones or PDAs (personal digital assistants). Writers such as Gillmor (2004) believe that the internet is improving journalism generally since it represents a challenge to corporate approaches.
Corporate news	Sometimes businesses set up blogs in order to, ultimately, promote their product. They may trace the daily life of the business or feature a range of different aspects of the business on a regular basis. Sometimes such blogs are not obviously part of a business—they may be written like a personal diary type blog and regularly refer to particular products in order to promote them. Many newspapers have blogs as additional features to their usual online publications—for example, *The Guardian* and the *New York Times* have blogs that in some ways are more informal and provide opportunities for readers to comment and interact with journalists.
Personal journals	These are perhaps the most well known of blog types, leading some people to believe that blogs are a form of journal. These blogs typically provide daily accounts of the blogger's lives. Many celebrities use this format as a way of promoting their image and activity, and indeed some politicians (e.g., David Cameron, leader of the UK Conservative party) have used the blog to promote a particular type of self. Often personal blogs have themes—such as life in war-torn Baghdad or, on a much lighter level, the chronicles of life as an airhostess.
Fiction / creative writing	Some blogs are written purely as creative pieces, and in role as characters. These might be based on existing works of fiction—such as *Wuthering Heights*—or on invented characters.
Fan sites	Some websites are totally dedicated to particular celebrities, football teams or sporting heroes. Alternatively they may cover a range of celebrities and feature gossip and images.
Gadget showcase	These blogs feature and review newly invented gadgets. They usually have multiple links—to, for example, other reviews, promotional websites or sites where they can be bought. It is clear that businesses often send in items for review and thus contribute to content.
Hobby / personal interest	There are a great number of blogs where the writers focus just on one topic—for example, fashion, cookery, photography, Second Life or knitting—the list is endless. These blogs might, for example, offer a daily recipe or a knitting pattern. Often the blogger shows what they have been doing each day in relation to their hobby. Sometimes such blogs are linked to other sites that the blogger is involved in. This kind of linking allows the person to reveal a wider online identity.

Figure 3.   Popular blog types

TYPE	EXAMPLES
Students	There are a whole range of blogs already on the internet where pupils have written as individuals on their own blogs or contributed individual posts to class/group blogs. These blogs might describe what pupils have been doing on a daily basis and give a review of their reflections on the work. They might be associated with particular projects and provide particular information on subjects the pupils have researched—examples include dinosaurs, Egyptian tombs and so on. Other blogs may involve pupils writing diaries of characters of books they are reading.
Teachers	An increasing number of teachers keep blogs about their use of digital technologies in teaching, about their daily teaching lives in general, about new ideas in their work and all matters relating to their professional lives.

Continued

As we have seen from our initial exploration, the blog format offers a range of interactive and collaborative possibilities for individuals and teams. Some of these possibilities derive from features that are part of the architecture of blogs. But it must also be recognised that during the last five years, a period in which the blogosphere has undergone a rapid expansion, diversification and innovation have been of central importance. So, for example, whilst admitting that blogs are an unstable form, as they continue to mutate and hybridise, Lankshear and Knobel (2006b) offer a provisional taxonomy of blogs identifying 15 different kinds of blog. There is clearly no standard way to blog. Arguably, the single defining feature of a blog is that of date-ordering (Walker, 2003). Although periodic updating is also a feature, this is contentious, because some bloggers post daily whilst others are less frequent. This in turn begs the question of how long can a blog be considered to be 'live'. A blog owner may decide to completely delete a blog or alternatively may gradually neglect it. In effect, a neglected blog is likely to attract fewer and fewer visitors—perhaps gradually disappearing as a feed or a favourite. So although updating is difficult to quantify, it is important in sustaining a readership, and, of course, readership is important.

Who's blogging?

Studies of the blogosphere include Mortensen and Walker's seminal early work (Mortensen and Walker, 2003), our own study of academic blogging (Davies and Merchant, 2007) and Lankshear and Knobel's (2006a) typology of blogs. These descriptive and analytical studies reflect the writers' own involvement with blogging. They also report on the wider use of blogs across the internet, providing theoretical accounts and understandings of blogging as a social practice and a new kind of literacy. However, data that covers the actual spread of blogging is a little harder to come by, as we shall see.

Reliable statistical information about any aspect of Web 2.0 use is not as easy to collect as we might think. This is partly because of the lack of stability caused by the continually shifting pattern of engagement with new technology, and partly because of inherent difficulties in capturing that information. In the case of blogging, it's well known that many novice bloggers set up and abandon blogs without ever adding to their first post, and seasoned bloggers may own several blogs—some of which they maintain, some of which they do not. It is difficult to define a 'live' blog or even to ascertain whether a blog has been truly abandoned or not. To make matters worse, headline-grabbing statistics are often hijacked by commercial service providers to promote and publicise the popularity of their own products or are used as ammunition by the critics or campaigners who fuel moral panics. Determining the demographics of those who use blogging software is further compounded by difficulties in reported use as opposed to actual, active use and the standard problem of identifying who users actually are. To say that there are more than 50 million live blogs (Technorati, 2006) or 30 million people on Facebook (Zuckerberg, 2007) provides us with only a very crude snapshot. There remain many who still have no idea what a blog is, while others participate in communities within which blogs (and other related software) play an important role.

It is very easy to create a new blog. Even with a quite basic knowledge of computing, you could set up a site in under five minutes. When we read that there are 150,000 new users of Facebook each day, it is easy to misinterpret the information. That sort of blog-count simply tells us how many times that five-minute joining operation has taken place. It doesn't tell us how many people have multiple blogs, how many were created for demonstration purposes or how many are regularly updated, regularly visited or commented upon.

The format of blogs

As mentioned above, trying to characterise blogs is probably easier in terms of describing their format rather than their semantic content. Although there are variations in format, blogs commonly share a range of features that make them quite easily recognisable as blogs.

A blog contains certain standard textual features that tend to 'frame' the changing text that accumulates as the blogger adds to his or her postings.

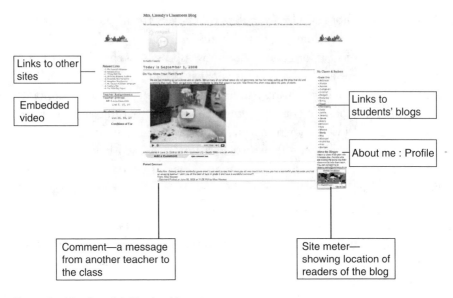

Figure 4.   Mrs Cassidy's Teacher blog.

Typically, blogs will have a title running across the top of the screen—sometimes this will also show the software used (such as Blogger, Wordpress, Xanga etc.). To either side, marginal space is provided to show links, pictures, details about the blogger and so on.

In this example of a blog (Figure 4), the most recent post, displayed centrally, is about a project on plants. There is a video embedded in the blog, showing one of the pupils identifying a plant part. The children contribute to the blog quite regularly and in a recent post, for example, took a series of images that have been put into a slide show about the letter 's'. The blog is thus jointly authored but remains under the organisational structure set by the teacher. To the left of the blog are links to other sites, one of which is also maintained by the teacher and acts as a portal to all the other sites she has set up. There are also links on the left to other teachers' blogs. To the right, is a links list that connects to the children's blog posts. Here they can describe their work and their reading activities. This blog works as a central point for displaying the children's work, acting as a kind of anchor for Mrs Cassidy's teaching. We have followed this blog for some time, and the teacher identifies for the pupils in her class as well as for outside readers what the class is doing, what they will be doing and why. It both reflects her pedagogy and is also a medium through which she teaches and the children learn. She is clear about the benefits of opening out the work to others, saying in her 'About Me' section, 'I teach a class of six year olds in Moose Jaw, Canada who are inviting the world into their classroom to help them learn'.

Through her use of links, Mrs Cassidy is also able make great use of the connectivity across the web. Donath and boyd (2004) suggest that 'public displays of connection' are a pervasive feature of social networking. These public displays are strategic in that they make public what the blogger's interests are and, therefore, useful in making links amongst people with similar interests. Bloggers frequently make use of these affordances, both to demonstrate connectivity (and raise their social profile) as well as to extend it; links are a way of showing allegiances as well as a way of forging new ones. Of course, this kind of strategy is now well known amongst young people who link with others through Facebook, MySpace and Bebo, for example, but in the case of education practitioners, this can be useful in that teachers can form supportive networks and share ideas.

Connectivity is useful to bloggers since it allows them to interact with a wider community. For example, if someone reads a blog post and wishes to comment, they can easily do so—by clicking on the word 'comment' that follows the posts in most blogs and then leaving a remark in the space that appears on the screen. In doing so, the commenter leaves a digital 'footprint' or signature, since they are usually requested to leave an email address and/or a link to their own blog. The signature or footprint left on a comment is immediately hyperlinked and so anyone reading the comment can 'click through' to the commenter's blog. This frequently leads to reciprocal blog reading and often the inclusion of frequent commenters on the visible 'blog roll' in a sidebar. It is easy to see how in this way, networks of similar blogs can quickly establish themselves as affinity spaces (Gee, 2004b). As a particular network grows, bloggers increase not only their readership and online presence but also their knowledge, or even social standing, in a particular area. It is easy to see, just looking at these few features, how the format or fabric of blogs contributes to social networking.

Other add-ons or 'widgets' (such as calendars, bookshelves and slide shows of photographs) involve a simple process of cutting and pasting in order to add them to a blog site In this way, a blogger can personalise her own blog, creating a space that can easily look like a professionally produced site. The more the awareness of other social networking sites, the easier it becomes to embed features and add textual materials to a blog. This is an attractive option, since the bloggers can quickly populate their blog with interesting tools and links, and in turn, social networking sites become ever more popular online spaces.

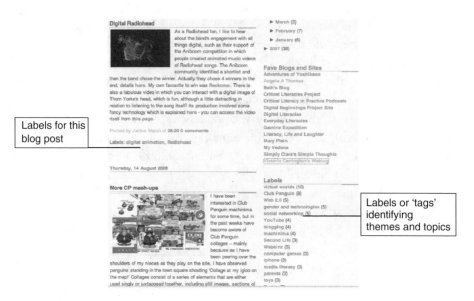

Figure 5.   Digital Beginnings: an academic blog.

Other blogs use a whole range of ways to establish links, such as Jackie Marsh's Digital Beginnings blog: This blog, authored by a literacy academic, allows her to communicate with other academics and her students about her research as it progresses. Not just within her posts but also in the sidebar links, she shares information about events and activities within the academic community. The 'Labels' section allows readers to explore themes and topics in the blog, since the author has labelled each one of her posts according to its topic or themes. Jackie Marsh sometimes tells of conferences she has attended, often linking to her papers or slide show presentations—both for those who attended the conferences as well as for those who did not. Hyperlinks within the posts themselves allow bloggers to 'point' to the organisations they refer to and literally embed one text within another so that information is quickly shared. This allows bloggers not only to disseminate their views but to also promote an academic presence in what is becoming an important space for debate—the blogosphere.

Postings can be of whatever length the writer wishes—although short, well-written and concise posts are more likely to be read to the end and may attract more commenters, the life blood of some blogs. It seems that for many bloggers, comments keep them going, with an awareness of audience inspiring them to write more. This is, of course, something many teachers are already aware of—a genuine and interested audience is more likely to inspire and engage writers.

Blogs as a learning tool

The sequential, chronological characteristics of the blog format suggest how it can be useful in capturing such things as the development of a narrative, the design and implementation of a project, the progress of research, emerging processes, the aggregation of links or references, and observations or reflections that develop over time. Blogs, as multimodal texts, also allow us to represent these activities in text form, in still and moving image or in audio format—and, of course, some of the most interesting blogs are a judicious combination of these modes. Educational blogging can capture learning as it unfolds over time and this has obvious benefits for both learners and teachers. In this most basic sense, a blog can provide an analytical record of learning or an online learning journal (Boud, 2001).

Writing in 2003, Efimova and Fielder noted that alongside the 'diary-like format' blogs kept for family and friends, there was a:

> growing cluster of weblogs used by professionals as personal knowledge repositories, learning journals or networking instruments. (Efimova and Fielder, 2004:1)

They go on to suggest that these newer blogs not only serve the needs and interests of those writing them but also display emerging ideas in a public space. This suggests the development of more open learning journals that can be interlinked and commented upon within an emerging community of learners.

As Richardson (2006) points out, blogging can also involve users in an important and distinctive kind of learning; one that he characterises as: read- write- think- and -link. Here he suggests that a blogger develops a kind of practice described as 'connective writing' in which active reading and involvement through comments and hyperlinks work alongside regular posting in the co-construction of meaning—or, to use the language of this book, learning through social participation. This view accentuates the significance of a community of bloggers, in the form of either a cluster of related blogs or a team blog. From this point of view we can see blogging as a way of supporting a community of practice (Wenger, 1998) or as an affinity space (Gee, 2004a).

Perhaps it is worth introducing a note of caution at this point—blogs in themselves cannot be seen as promoting new kinds of learning through social participation, nor do they necessarily involve new literacy practices—it is more the case that through their social and technical affordances they present opportunities for this kind of work. In planning classroom work, we need to be sensitive to these distinctions and to think carefully about the pedagogic implications of using Web 2.0 technologies.

In Vignette 2, given below, the teacher has considered a number of ways in which blogs can be used to support learning. Early in the project, blog teams are using the medium to express personal opinions and to get feedback through comments. This reflects mature, everyday blogging practice, albeit under the direct guidance of the teacher. She also encourages her pupils to use their blogs as a repository of information through the use of hyperlinks, whereas documenting the class field trip helps pupils to explore the visual affordances of blogging. The blogs have a reflective quality and because they are internet-based, they can also be accessed by parents. Clearly these blogs provide a context for collaborative work, both within and between the teams. However, they do not attempt to create opportunities for wider social networking—this is beyond the scope of this particular project.

> Miss Gupta's class of eleven-year-olds is investigating river pollution as part of a unit of work in Geography. She has organised the class into mixed-ability teams each of which will keep a blog. The initial posts are used for statements of what each team thinks about pollution and the environment. After she has taught them how to insert hyperlinks, they are encouraged to search and evaluate web-based sources, recording and commenting on these in their posts. Later, on a field visit, they take digital photographs of environmental hazards such as fly-tipping, invasive non-native flora, industrial effluent and so on and then upload them onto their blogs. Towards the end of this unit of work, pupils are asked to reflect on what they have learnt. Miss Gupta sends a letter home with the blog addresses inviting parents to visit and comment if they wish.
>
> Vignette 2—Blogging river pollution

By contrast, the second example (Vignette 3) explicitly aims to create new connections between students. For sound pedagogical reasons, this has to be carefully orchestrated by the two teachers involved (and this incidentally brings new forms of professional collaboration into play), but peer-to-peer interaction and learning are a central feature of these blogs. In this project, evaluative comment and feedback are key to its success, and background work on appropriate ways of doing this provide an important support. The action-adventure blogging relates to 'real-world' practices and students were able to use popular blogs such as those at 'Bill's Movie Reviews' at www.http://billsaysthis.com/movies and 'Independent Film Review' at www.http://moviereviewblog.net for ideas.

Both of these examples show how classroom projects can use the affordances of blogs to support learning. Although the blogs concerned have the potential to attract a wider readership, they were designed to serve quite specific and time-limited purposes. Other instances such as photo-blogs chronicling the progress of a new school building or charting the fortunes, fixtures and fitness of the school football team have a longer shelf-life and may attract a more varied readership. Nevertheless the vignettes presented show authentic blogging practices (i.e., blogging that reflects everyday or 'real-world' uses of blogs) that are carefully interwoven with classroom learning.

> *Students in a rural school have established a partnership with a similar-aged class at another school several hundred miles away. Both have been studying the action-adventure genre using film, comic and book sources. The two schools use class blogs to exchange and comment on each others' views and opinions. The work begins with posts about favourite films (and tie-in video games such as 'The Golden Compass') and then becomes more focused as the students discuss action sequences and favourite characters. They later begin to post plans of their own stories. When the stories have been written, the students make short video trailers that are then posted on the blogs. Requests for the full stories are made in blog comments. The stories are then exchanged by email and 'review' posts are displayed on class blogs.*
>
> Vignette 3—Action adventure blogging

Blogging for teachers and pupils

In our work with teachers who are using blogs in an educational context, we have watched them take tentative first steps, sensibly aware of the contentious nature of connectivity and keen to ensure pupils' safety. They have often taken care to protect pupil identities and have not encouraged pupils to leave comments on blogs outside of their own projects. Some are beginning to see how blogging can transform learning in their classroom, providing connections to learners in other contexts, to experts in different settings as well as to their own families and communities.

Although many teachers are keen to use new technologies in school, we argue that ideally, as their confidence and expertise increase, they should recognise the real value of using the affordances offered by connectivity beyond the classroom. Such experiences with the broader blogosphere will allow learners, under the guidance of teachers, to explore the internet in safe ways and learn to read and critically interact with the vast amount of text now available online.

Duffy and Bruns argue that both teachers and learners need to keep up with new developments in digital communication:

> The uses of these technologies—and the technologies themselves—are still developing rapidly, and teachers as well as learners need to keep track of new tools and approaches

emerging both in the wider internet community and specifically in educational contexts. (Duffy and Bruns, 2007)

Blogs are now a well-established and widely recognised form of digital communication, and this alone suggests that they should be taken seriously in educational
settings. As we have argued, the software is easy to use and can be particularly
motivating for learners to use. In blogging projects, children and older students
can gain first-hand experience and learn about the new literacies. They can learn
how to use the affordances of different modes in making meaning and how to
hyperlink their text to others, and to get a sense of what online participation is
actually like through commenting and responding to comments, on their own
and through other people's blogs. We have also argued that the blog format offers
new and different ways of capturing and presenting learning in a purposeful way.
And, for the more adventurous, the blogosphere can transform pedagogy by
opening up classroom learning to a wider, dispersed audience.

· 4 ·

PHOTO SHARING—WELCOME TO THE WORLD OF FLICKR

Flickr is one of the most popular online photo-sharing sites, which—according to its own blog—had, by November 2007, hosted more than 2 billion images (Flickrblog, 2007). At the moment of writing (20 August 2008—5 pm, UK time), 5,500 images per second were being uploaded. Launched in 2004, and arising initially from a multiuser game template, it was one of the first Web 2.0 applications (Davies, 2006) to achieve popular appeal and is one of the most enduring to date. At face value, it is a site to which users can upload images for others to see and comment on; yet as we have found with most Web 2.0 sites, it is far more than this and is constantly in a state of renewal and development as users and official Flickr developers continue to push at the boundaries of what is possible.

This chapter draws on our own research and experience in this environment, which, as for many other users, began because we needed somewhere to store images to be used on our blogs. Flickr is a site that is used by many people for many purposes; the notion of photo sharing may not at first seem to be of interest to many people other than photographers, however, photography is only one of the many topics discussed on the site and is not the primary interest of the majority of its members or 'flickrites'. In this chapter, we begin by exploring some of the key features of Flickr, before going on to look at its social networking features. We show how these can provide a context for both formal and informal learning. We also look at ways in which online interaction occurs around images, and at how the kinds of image that are placed on the site are affected by other images, by online activities and games organised through groups, by discussion threads associated with groups and by comments, titles and tags attached to specific images. We also explore ways in which individuals choose to represent their lives and aspects of themselves and reflect on the ways in which this com-pares to blogging.

Photo sharing with Flickr

Flickr provides a service that allows for the online storage of digital photographs and, more recently, videos. One can use the site without joining, and it is possible to search for images, either via the sites own search facility or indeed via other search engines such as Google or Yahoo. Exploring the site allows visitors to look at single images in a range of sizes, to see comments on the images, to look at users' photostreams, to see images within the context of sets made by individu-als, or to look at groups of images contributed to by many. Registering at the site immediately opens up additional functionality, so that one can upload images, label and title them, comment on others', set up groups, complete a profile and 'friend' others.

Figure 6 illustrates the homepage or portal interface of 'on-the-run's' flickrstream and gives an impression of the Flickr look. On this you can see recent photographs, how sets can be organised, and general navigational.

If one is a member, the homepage is one's own Flickrstream; non-members see the Flickr welcome page that invites readers to join, to search the site, or 'take a tour'. As with most Web 2.0 sites, the design incorporates prompts in many spaces for encouraging membership, including a range of ways for inducting visitors into how to use the site and understand the rules or 'community values'. Participation is key to the survival of such sites, even though membership is free—once members get involved, they invariably want to do more than the basics and so wish to take up additional storage space and increased functionality (extras that come at a premium). Thus enticement to become 'socially' involved begins on the homepage, with an invitation to search the site, and later to join groups or make comments on images.

Figure 6. The Flickr portal.

At entry level, once visitors have signed up, they can develop their own pages that become a portal through which they can upload and organise their own images. As with blogging, the most recent items appear at the top of the page, with a date and a time stamp. However, it is also possible to organise one's images into sets, and sets into collections of sets. Thus one might organise a series of images into a set called 'family'; this may then be placed in a collection that might, for example, include sets named 'portraits', 'people I work with' and so on. The capacity of Flickr to organise images in many ways is one of the features much praised by its members, and over time additional features have been added so that members can make use of a broad range of permutations. This facility allows images to be displayed in many ways, so that a single image can be viewed in many contexts and be given a range of meanings according to the different contexts it might find itself in. Putting a series of pictures into a set has obvious educational benefits as the vignette provided below illustrates.

In a primary school, Ms Hooper was working with her class on a science project looking at the growth of plants from seeds. In groups, pupils each grew different types of flower and, using digital cameras, they recorded the process of planting and growth over the coming months. Images were uploaded by the pupils to the private Flickr account set up by the teacher for the project. The groups arranged the images in sets according to a range of criteria. For example, the 'sunflower group' had its own set, as did the nasturtiums and pansies. Another set was called 'seeds', others were 'seedlings at one week', 'seedlings at

two weeks' (etc.) and yet another 'adult flowers'. Reflecting on sets labelled and arranged in
this way helped the children to see the images in terms of illustrating how flowers grow and
develop in comparison to each other. In addition to arranging images in particular ways so
that additional meanings could be located across and between images as well as within them,
the children used Flickr's notes facility to superimpose textual notes to images so that specific
aspects of the sunflower (for example) could be labelled. The children gave each image a
title and wrote descriptions about what they could see. Each time they were asked to try and
capture the changes that were taking place.

Vignette 4—Seed photography

This vignette describes how one teacher used Flickr in her classroom. The pupils were excited to take photographs of the processes and could enjoy each other's work, by commenting on images belonging to specific groups. The way language was used alongside the images and sets as well as how items were framed by the children's shots made the botanical points more or less clear—so that children were able to understand the skills relating to visual presentation as part of text making—and their photography improved. Naturally literacy learning was embedded in the tasks as students labelled, described, titled and commented on the images. In terms of text creation, it is easy to see how Flickr illustrates the ability of linguistic context to impact on textual meanings, as well as how other images work alongside others for comparison and contrast purposes. This example indicates ways in which Flickr can be relevantly utilised within classroom activities—even before we begin discussing interactive opportunities that are offered when the images are made accessible to others beyond the classroom.

The Figure 7 illustrates how an image on Flickr is surrounded by written text (Dr Joolz, 2007). This text is organised by the Flickr template, but the actual words are chosen by users. Our example shows how Flickr lists sets, how it provides a space for a title, a description and comments. The photograph in the figure was taken in New York City; it is a public noticeboard that has had all of its notices removed. The remaining scraps of paper have created a mosaic effect of many tiny scraps of paper that attracted the photographer's eye. The image has been placed in many sets that are placed in a hyperlinked list to the right of the image. Any viewer can see the image in the context of those sets by clicking on the hyperlinks.

Also to be seen in this figure are the hyperlinked tags associated with the image: for example, 'scraps', 'NYC', 'palimpsest', 'paper', 'text', 'staples'. Evidently the tags refer to the place where the image was taken, as well as to content and aspects of the image. An important feature that Flickr shares with other SNSs is the use of labels or category tags. The tags can be used to search within one's

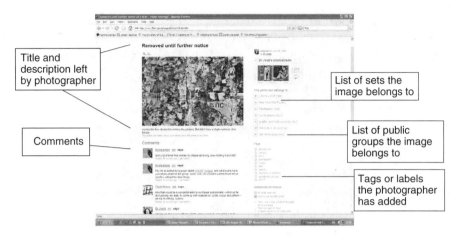

Figure 7. Flickr sets, comments and tags (Dr Joolz, 2007).

own stream or in everyone's photos. Clicking on a tag such as 'palimpsest' will conjure all images that have been tagged with the word 'palimpsest'. Thus a tag search will change the context of individual images, showing an image alongside others tagged in the same way and thus reflecting something about the meaning of the term 'palimpsest' and perhaps even developing a particular meaning of that word for Flickr photographers. Category tagging is a process by which objects or ideas can be classified. So in blogging, photo-sharing and music-sharing sites, one can code content with keywords or tags that can then be searched for and grouped in a variety of ways. Of course, in some ways there is little difference between this activity and the established academic practice of attaching keywords to journal articles; however, there are some small but significant differences.

First, category-tags in online social networks are primarily generated by user interest, rather than by pre-set norms and conventions. Second, category-tags can be changed, updated or added to as new relationships to other objects are realised. And third, other people can add category-tags to your objects. This allows objects to be pooled and grouped in diverse and fluid ways in a process that is controlled by the community of users, rather than an elite group. Users' values, interests and priorities are the ones that count in this kind of folksonomy, and these will change over time as the nature of the people and images continue to change. A folksonomy is responsive to change in data and interactivity. Imagine a library in which books and journals could be organised and reorganised at the click of a finger by subject, by topic, by date or by size and colour—or whatever category readers apply—and you begin to understand the magic of a folksonomy.

In the example shown in Figure 6, not only is the image tagged with the term 'palimpsest' but there is a set and a group of the same name to which the image belongs. Further, one of the commenters, Clydehouse (2007) refers to the nature of the palimpsest in the image—albeit he mentions 'traces':

> now that would be a wonderful entry to our traces assignment—which so far isn't proving very easy to come up with material for. Lovely colour and pattern—art out of nothing.

The comment illustrates the way in which an obscure concept can be brought to the fore through tagging and interactivity—and certainly, as we mention above, it shows how particular concepts can develop in particular ways within affinity spaces or 'Communities of Practice'.

Interestingly, the Flickr administration displays a 'tagcloud' to reflect back to its users the nature of the folksonomy it has collectively generated (Figure 8). Such a display seems to us not just to function simply as an innocuous illustration of users' tags. The tagcloud has a function that goes beyond the mere recording and presentation of users' tagging activities; further it does not simply respond to users' values. The tagcloud has a further role in affecting behaviour- encouraging users to develop an awareness of the whole community of users, the way it is developing, and to reflect upon one's own participation within it. At a basic level, it can give ideas of things to look at and photograph, but further it plays a role in facilitating a sense of community. The endorsement and encouragement of self-conscious participation with others, awareness of involvement in and responsibility for its value system, seems ubiquitous across a whole range of social networking spaces—such as in the eBay discussion forums (Davies, 2008), YouTube's reporting of 'inappropriate content' facility, and a plethora of feedback possibilities on Flickr.

To join Flickr, as with any other social networking site, is to become part of a community. However, users determine their position in the community by their level of engagement with others. In other words, you choose your own level of participation in the larger community. So for some users, the central motivation for using the service is as a way of storing photographs on someone else's server, perhaps in order to make space on a hard drive, or to serve as a back-up. These may be stored in a way that allows public viewing access, or they can be made totally private. Alternatively, users can allow contacts, friends, family and 'temporary guests' differential access to these images—so that family members may be able to see additional photos that mere contacts may not, or that friends can see images that the general public cannot. Controlling access in this way allows

Figure 8. The Flickr tag cloud, showing most popular category tags.

different types of interactivity level over different types of image, and it means that individuals can feel confident about privacy. This kind of security means that the space can be used by teachers and pupils in a range of ways according to levels of confidence and appropriateness to the task. Thus images could be opened out for viewing by parents, groups of pupils in the same or different schools, or by other trusted agencies.

As with many sites, members of Flickr are often keen to attract many visits to their photostream and may use a number of tactics to encourage such visits. As mentioned above, Flickr is designed to optimise interactivity not just with content but also with users. These two aspects work closely together, and many Flickr members become very motivated by interaction and participation. One method is to encourage others to their photostream by using long tag lists, so that multiple searches will often find their images; another way is to set up groups and invite specific others to join; another way is to comment on many other people's images—who are then likely to reciprocate. Further, Flickrites can 'friend' people, by simply adding others to their list of contacts, friends or family. Once identified as having been selected as a friend or as a similar contact, other members will receive an automated message from Flickr, saying who has 'friended' them and including a link to the initiator's photostream. Flickr thus uses a broad range of strategies to encourage interactivity and this process helps to promote the site and increase patronage. Despite this clear objective for business success for Flickr, the site nevertheless provides a wealth of opportunities for educators and learners.

When we each joined the site in 2004, it was much less developed than at present, with far fewer members and far fewer features to allow for the person-alisation of photostreams, less choice about levels of interactivity with other user groups (friends, family, public, 'guests' etc.) and so on. However as membership increased, Flickr responded to ways in which users were behaving, as well as to specific requests via discussion boards and email—a participatory design approach. Functionality increased, as did user participation, and membership began to increase phenomenally—the site gradually developed so that emphasis for many users shifted from utility to a space where a whole range of different types of activity could occur and where awareness of other lives, other cultures and other ways of seeing the world opened out (Davies, 2006).

Guy uses Flickr principally as a place to store images for his own blog. Although he understands the concept of tagging, uses it, and understands its role in creating folksonomies, his interest stops there. He is not particularly interested in establishing much of an online identity in the Flickr community. On the other hand, Julia's interest, like that of many others, developed from her original use of the space as a repository for her blogging images, and for a long time she concentrated on text making and participation in Flickr far more than she worked on her blog.

For Julia, a key way in which she was enticed to become interactive on the site was through participation in public groups. Public groups can be set up by any member, who can invite others to join and contribute to an image collection. Group 'admins' can not only set up the groups but also adjudicate on membership rights, vet images and set up rules for participation. There are many totally private groups on Flickr so that some activities occur unwitnessed by non-members, and whose tags do not appear on the ever-changing Flickr tag-cloud. As shown earlier in Vignette 4, groups might involve membership amongst only those who already know each other and who see each other daily and, as in the vignette, may be in the same room as all other group members even at the time of uploading their images. However, many groups comprise only those who met in the Flickr space and who, therefore, form relationships through the images they show and the text they create around them. In this case, the images and text become the way in which the users represent themselves, so that interactivity requires greater care for communication to be successful, safe and clear. The choice of group to join and the type of image to share become key identifying factors in the presentation of self and provide a starting point for interactivity with others.

Many Flickr members collaborate intensively in groups, where they not only pool their images in such spaces but also open discussions about those images, the circumstances under which they were taken and how they fit (or not) the group's definitions. Discussion can remain steadfastly about the images and content of the group—but frequently interactivity develops in such ways that identities are explored and presented through the modalities of word and image. Interactivity is usually enthusiastic and lively; people learn about each other's lives—often allowing for cross-cultural comparison and learning; mentoring relationships often develop; in-jokes emerge through banter and fun; people sometimes even email or send gifts; and it is often through groups that new friendships might form that result in face-to-face interaction.

Some groups are specifically about teaching new skills and provide work-shops on digital image manipulation, yet most are not about photography per se. Groups may comprise collections of images that feature specific colours, shapes or perspectives, such as: 'The Red makes it', 'Beautiful Green', 'Squared circle', 'Looking down' and 'Shooting up'. The existence of such groups certainly suggest new ways of grouping photographs, but, as Julia has found, they also influence the kinds of photographs she takes and ultimately led her to look for particular photo opportunities. It was through Flickr, for example, that Julia became interested in street-art—firstly noticing the images online and then starting to really look at examples offline, on the streets in her locality. This interest took her to other cities, even other countries, wanting to see for herself (and to photograph) images that she had seen only online before. Over time, Julia learned from others online about different styles of art, different artists and different techniques—thus developing her own preferences. In this way, Julia found that, like many others heavily involved in Web 2.0, her participation in online communities had an effect on the way she behaved offline and influenced the way she saw the world.

The case of the padlock group

In the summer of 2006, Guy's engagement in a new Web 2.0 research project—which involved the creation of a virtual world—led to a more intensive use of Flickr. He needed to collect photographs of walls and doorways to use as textures in the construction of the virtual world. He uploaded these images to his Flickr photostream and, to allow easy access by the designers, based in Finland and the USA, stored them in a set called 'Pieces for a virtual world'. One such image was a shiny silver padlock against a strong blue background (Figure 9). Uploading this image, he decided against 'dignifying the everyday'—which he often used as a tag—and went for the more obvious 'padlock' descriptor. Within a matter of hours, he had been invited to join the padlocks group, a micro-community within Flickr that specialises in padlocks!

Evidently Guy's image had been located by the 'padlock admin' via a tag search—such was the motivation of the admin to keep the group dynamic and comprehensive. It is at precisely this point that the business of category tagging becomes a significant social practice of literacy in its own right. In this instance, the tag becomes a gateway to what Gee (2004a) describes as an affinity space. Affinity spaces are described by Gee as being guided by purpose, interest and content. Thus the endeavour or interest around which the space is organised is for Gee the primary affinity; it is less about interpersonal relationships and more about the exchange of information itself.

Having accepted the invitation, Guy had now taken on, as far as the affinity space was concerned, a temporary identity as someone interested in padlocks—

Figure 9. The original padlock image with category-tags.

not a piece of information he was particularly keen on sharing with friends—but nevertheless this was a new kind of engagement for Guy!

As with Julia and her new interest in street-art, this engagement with the padlocks group alerted Guy to an aspect of the urban environment that he had not previously noticed. He began to look and think about the ways in which we are locked in and locked out of certain spaces. In short, he began looking at his environment in a new way. As he posted his pictures, he also became aware of the different makes, sizes and ages of the locks, their serial numbers and so on. The process of categorisation led to the accumulation of new information as well as a new way of seeing. But, of course, the collection of information, the sense making, the organisation of information through categorisation and the trading of detail and knowledge describe some of the essential processes of human intelligence.

Admins of some groups are very stringent in ensuring appropriate participation—they set up very specific criteria for contributing images and watch carefully to ensure all images fit the rubric. Members are, therefore, required to think carefully about the taking of their photographs, and how to use language to help them fit a particular group. The members need to use their observation skills to read the texts already in the group, and to be able to find and appropriately frame images to fit. In terms of learning, participants start to understand how the framing of a particular image may give it a slightly different nuance; how perhaps the manipulation of colour (maybe through lighting, type of film or even through the use of software such as Photoshop) can affect the meaning of an image. Groups are an excellent way of structuring activities and, whilst allowing a whole range of creative responses, provide rules for participation and learning. Possibilities for educators here are multiple—with opportunities for either teachers or pupils to set up groups with particular aims and objectives. These might relate to a whole range of curriculum interests, from geography and English to mathematics and the sciences.

For example, substitute the attentive noticing of padlocks for looking at squares and circles and you have a familiar item in an early-years mathematics curriculum; images of the locality might identify particular problems for a social geography project; categorise lifeforms into vertebrates and invertebrates and you have a fundamental building block for the natural sciences. In this way, one could argue that category tagging and building folksonomies has an important role to play in illustrating knowledge-building practices between dispersed individuals and also shows how a new form of digital literacy has educational implications.

Modelling the process

One of the fundamental features of the example of category tagging we have given is the way in which it is socially located. After all, Flickr is described as a social networking site. As we have shown, the degree to which the user invests time in networking can be hugely variable. However, it is only through social participation that one becomes part of a group. If the sharing of interests through the pictures we take comes close to a kind of learning (and maybe there are some situations in which this is more central than others), it quite clearly constitutes socially situated learning (Wenger, 1998)—despite the fact that face-to-face contact between participants is unnecessary—although this is hugely variable too. In passing, it is worth noting that some users interact only with those who they already know offline, while for others, such as Julia, Flickr opens out whole new vistas of entirely new contacts and relationships.

Another feature that seems important here is the use of literacy—that is literacy in the sense of 'lettered representation'. This is interesting, because Flickr is driven by the visual image, the digital photograph being the social object around which networks take shape. But words, rather than images, are just more useful in categorising, or in giving titles, descriptions and feedback. As Julia found in a school-based project on critical literacy with a group of fourteen-year-old pupils, the meanings of images could be changed dramatically via the application of different titles, tags and descriptions. Further, over time, if images accumulate a series of comments, the image comes to mean something different to those involved in the discussions (Davies, 2007). Additionally, verbal notes can be added to images, superimposed on the image itself. Such notes might merely label but could also suggest new ways of interpreting aspects of the image or highlight an inconspicuous detail.

Other literacy activities on Flickr include rubric writing for groups, discussions in groups about images and their inferences, and, of course, when one Flickrite wishes to invite another to join a specific group. Icons and automated messages can now fulfil the group invitation function, but we argue that the affordances of this kind of digital literacy make it a powerful tool for organisation and interaction and actually, when we view Flickr as a whole, provide the mechanism for the sifting and flowing of visual images.

In Figure 10, we have modelled the process of what happens when we start category tagging in the ways described above. In the first part of the cycle, we distinguish between the everyday sense in which we see the world and what we refer to as *attentive noticing*. Seeing can transform into this attentive noticing when

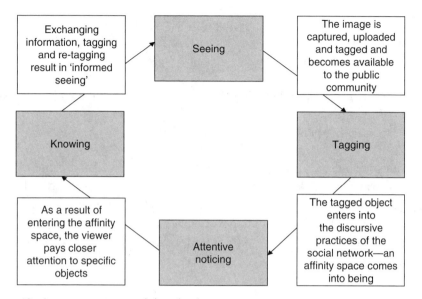

Figure 10. Attentive noticing and the role of category tagging.

we begin to label things in our environment. This act of labelling is normally linguistic. It could be an oral or symbolic representation, but in the padlocks and palimpsest examples, the tags are in a written form. To suggest that the simple act of attentive noticing leads automatically to *knowing* is, of course, oversimplifying complex issues. It might be better to cautiously suggest that attentive noticing sets up the conditions for knowing. More importantly though, we want to argue that this cycle of events begins to transform our seeing into *informed seeing*, as we begin to look more closely. In Guy's case, beginning to distinguish between the Abus and the Yale seemed like an essential rite of passage for a padlock collector!

A final and important element in this process hinges on motivation and purpose. It seems to us that the amount of energy and resource that one is prepared to invest in a particular act of knowledge building will determine the level of social participation, and the learning that takes place—in short, the degree to which one identifies with the affinity space.

Taking Flickr further

As noted above, there are many different types of public groups on Flickr that invite a whole range of activity. We suggest that teachers might set up groups seeking pupils' contribution or ask pupils to set up groups of their own. There are already some excellent ideas on Flickr that teachers may like to learn from, such as those that follow perhaps a particular artistic tradition:

- Martin Parr we UBoring PostcardsPhotograph like paintingDiane Arbus
- Name that Film.

Here, members demonstrate their understanding of a particular genre or style and build on previous knowledge. Viewing these images collectively enhances such knowledge and provides a particular perspective on many aspects of our world. Other groups focus on finding particular patterns such as repeated designs or marks, interesting prints or textures, reflections or symmetrical designs and so on. Many groups use the internet as a kind of virtual gallery for street-art or 'graffiti', with groups such as

- Walls Speak to usGirls on Walls
- Visual Resistance
- Stencils
- BanksyWet shame graffiti

Such groups provide a new context for showing street-art, bringing it a new lease of life in a new exhibition space allowing it to be seen less as vandalism and more as being part of a sociopolitical movement. These groups provide potential discussion material about social issues and politics and about the artists, and around images of street-art—whether it is a democratic art form or urban crime, for example. These are areas that would interest many pupils, and following discussions and participation would be an exciting endeavour for many.

There are groups that draw on games whose roots reach back to other traditions, such as

- Visual Bingo
- SnapPicture Dominoes

Such groups develop and require team work, collaboration and close observation skills for participation. Other groups have an interest in narrative, for example, the '5 picture story' group requires individuals to upload five images that tell a story; 'photo dominoes' requires members to contribute an image that relates in some way to the previous one—thematically or content-wise; 'domestic spaces—human spaces' requires images of items in domestic spaces that leave a clue of what has happened before.

Pupils can be involved in thinking about titling images, tagging them, offering pithy descriptions and commenting on others' photographs. Such activities allow students to write within templates and to think carefully about their use of language to highlight particular aspects of the visual text. English teachers might want to ask classes to take images that illustrate metaphors of their own making; or they may ask pupils to take images that illustrate aspects of a poem or play. Drama teachers might ask pupils to take images that represent moments in a play. Some uses of Flickr allow pupils to think in detail about the relationship between the visual and the linguistic. Across the curriculum, there will be other kinds of use, such as images of the locality used in geography (about which we provide further detail in Chapter 8, in relation to wikis); while for photography classes, the possibilities for encouraging peer review is immense.

> A group of students in their final year of school were asked before they left whether they would post some images to a Flickr stream that a teacher set up. They posted images that depicted 'Good Memories' of their time in the school. Those same students were asked to contribute to another set of images on that same photostream, when they left to go to university. Each student contributed to his/her own set within the stream. They posted images of life in their first months at university—images of social life as well as their academic activities and where they were living. Younger pupils at the school were invited to ask questions and make comments on the images as well as to start discussions if they wished in the associated discussion area.

> Often discussions would move beyond the image itself and move into associated topics that allowed younger students to air their concerns and fears in a non face-to-face context. The younger students found they were less embarrassed to air these concerns in the online space that felt both intimate, yet not too close, that is, they were able to communicate with others who they knew but who were not immediately present. They could spend time phrasing and rephrasing questions and not submitting the message until they were comfortable with the text. The images used were, therefore, useful springboards for discussion that went into deeper issues if students wanted to take up such an opportunity.

> Vignette 5—Communicating across transition phases

The potential of Web 2.0 for keeping in touch with others when in a different geographical space has been well utilised by individuals in informal, out of school networks. For example, the use of blogs or sites such as Facebook when young people take a 'year out' abroad is well known. Similarly families can display images for their loved ones in Flickr streams. This was a useful way of utilising Web 2.0 in a slightly more formal way but one that allowed students to chat in the kind of arena where they felt more comfortable—nevertheless, it was not quite as informal as, say, Facebook, and so the teachers did not feel this was 'risky'.

> Mr Baker was working with his Year 10 English class (14- and 15-year-olds) in a critical literacy project. The pupils had spent time analysing texts using a critical literacy framework and the project then moved onto the use of multimodal and digital texts and, later, Web 2.0. Students analysed websites that advertised a range of foods—including MacDonald's, as well as looking at healthy-eating sites designed to promote nutrition awareness. The pupils keenly analysed the way in which the producers of such texts used words and images to influence the readers. As part of this, they experimented with taking digital photographs of food. They learned about how to frame shots differently in order to imbue the images with positive or negative connotations; they could see how the placement of food in different contexts also gave new meanings to the foods—for example, the background of a very clean commercial kitchen gave new meanings to a sandwich, than that of a gutter outside the school! In addition to this, the children experimented with Photoshop, so that they changed the images from colour to monochrome; they made colours brighter or muter. They began to see how a single shot could be manipulated visually to provide different meanings for the viewer. These same images were then uploaded to Flickr multiple times, and different students added text to the shots. They were able to see how different titles, descriptors and tags gave the images different meanings again. Commenting on each other's work, pupils mentioned how surprised they were at how they could present the images in different ways by using language differently. Overall this was a successful project, albeit the teacher still felt unable to open out the images for public viewing and participation because of concerns that the project would move out of his control.
>
> Vignette 6—Critical Literacy

The project in this vignette, like the one before it, utilised some of the social networking aspects of Flickr despite keeping it within a controlled participatory group. Many teachers have used Web 2.0 spaces to extend and explore existing curriculum interests and this has allowed their students to learn from each other and to collaborate in new ways. We see further potential for projects such as this in involving other schools and other classes but believe that those teachers who open out their work still further reap exciting rewards.

There is evidence of teachers collaborating within Flickr on such groups as the New York City writing project (http://www.flickr.com/groups/nycwp/), Gotham Schools (http://www.flickr.com/groups/gothamschools/) and Classroom Displays (http://www.flickr.com/groups/classrmdisplays/). In such groups, teachers can share ideas for teaching and learning using the medium of Flickr to think about it as a pedagogical tool.

Within this chapter, we have considered ways in which Flickr meets our criteria of a Web 2.0 space in terms of the four areas of Presence, Modification,

User-Generated Content and Social Participation. We have provided examples of ways in which teachers have been able to use Flickr in ways that allow them to Modify their own content as well as that of others; how to generate content—both images and text—in meaningful ways; and how social participation can change the meanings of that content as well as enable learning. It is the aspect of Presence in which many teachers have less confidence, preferring to keep work within restricted groups and maintaining control over who participates. By gradually gaining confidence in the tools themselves and learning more about Web 2.0, some teachers are nevertheless finding that they can introduce their pupils to the wider networking possibilities of online spaces. Confidence to do so may derive from teachers initially reaching out to each other in these spaces, before opening out that possibility to their learners.

· 5 ·

YOUTUBE AS VERB . . . ITUBE?
WETUBE? THEYTUBE? . . .

YouTube is a video-sharing site; although there are quite a number of video sharing sites available, with Wikipedia listing over fifty user-generated sites, including Blinkx, Crackle, Flickr, GodTube, Google Video, OneWorldTV, Ourmedia, Pandora tv, Peekvid.com, Photobucket, Vimeo, VideoJug and Yahoo! Video. However, YouTube remains the most popular and most well known. Launched in 2005, it was sold by its original owners and was taken over by Google in 2006–to much consternation and concern over what might happen to copyright issues and content control (Carroll, 2008; Trier, 2007a). The site has quickly accumulated an impressive number of users from across the globe, appealing to individuals across age groups and apparently used in equal number by men and women, boys and girls. An impressive ten hours worth of video is uploaded to YouTube every minute. The user-base is described by YouTube as being between 18 and 55 and 'spanning all geographies' (YouTube, 2008a)–although, of course, this latter point is a moot one, since access is determined by internet availability and this is clearly not yet completely ubiquitous. On first sight, it looks like everyone is already somehow involved.

The site describes itself in this way:

> Everyone can watch videos on YouTube. People can see first-hand accounts of current events, find videos about their hobbies and interests, and discover the quirky and unusual. As more people capture special moments on video, YouTube is empowering them to become the broadcasters of tomorrow. (YouTube, 2008a)

The text promotes all possibilities here, with both personal as well as community-minded interests being served. After just a few minutes of perusal of the space, visitors to the site will quickly find a huge array of material, organised by the YouTube template with tabs across the top of the screen leading to new spaces called: Home, Video, Channels and Community. The first of these takes visitors to the YouTube welcome page; members arrive at a page that remembers their favourites, sets out their 'recommended' videos, gives information about any messages they might have from other members and so on. This page is about presence, whereas the Video and Channels are much more about User-Generated Content and Modification, with participation in content generation possible–but not essential–and then the Community area sets out the rules, regulations, help on how to use the site as well as discussion areas for new ideas and so on. It is a complex and busy site and each of the four portals leads the way to a whole range of new choices of paths to take. Uninitiated users need to spend some time looking to see what the site can offer them as the choices are so many. We found that we quickly came across content we found offensive, and yet navigating is easy once you are acclimatised and have found routes through where one never needs to encounter such material. As we discuss later, we recommend to all teachers that they negotiate their own paths around YouTube before using it with classes and that they embed videos from YouTube into blog spaces created to serve material to younger students.

    Much of the footage on the YouTube is created especially for sharing on that site and only there; many users make multiple visits during the day in order to watch and talk about others' videos, as well as to monitor the activity around their own uploads. Some of YouTube's videos were created by amateurs originally for other audiences (e.g., family holiday films to show other family members) but later added to YouTube; some of it is professionally produced material–such as publicity material; and some of it is footage that is professionally produced but captured (often illegally, against copyright) and placed on the internet for others. This category attracts a vast audience and comprises such things as music video, excerpts from a broad range of television programmes and so on. A very popular use of professionally produced footage is to produce new video artefacts, known as 'mash-ups' or 're-mixes' (Lankshear and Knobel, 2006a), which we define and discuss later. As mentioned in Chapter 4, Flickr users often use Flickr as a repository for images for their blog, and similarly YouTube sometimes acts simply as a host for video material destined for other sites–probably most often for people's 'Vlogs' ('Video-logs'). In this way, we can see the site as a potential affinity space (Gee, 2004a; 2004b) where users make different types of contribution that are valued in different ways.

Users of the site can choose to:

1. Watch videos
2. Respond to videos by leaving a written comment or rating
3. Respond to videos by uploading another video
4. Watch videos and report them as offensive
5. Select any number of videos and embed them into other sites such as blogs and wikis
6. Upload videos to the site–making them public, private or available to selected YouTube members
7. Subscribe to particular users' videos and be alerted each time that user uploads a new video
8. Participate in 'Test Tube' and upload new, experimental (alpha) video software and YouTube applications
9. Produce profiles of themselves, look at others' profiles
10. 'Friend' people and send email via YouTube

A really useful facility for teachers is the ability to embed a video into another site—this can be done simply by copying and pasting the video code into another online space. The transfer might be made, for example, to a closed virtual learning environment, to a blog or to a wiki. By transferring material teachers can provide students with an opportunity to discuss the video away from the YouTube site and perhaps surrounded by other related materials that they have selected.

Use of the site is totally free; anyone can watch the public video material. Once you register, you can access the applications that enable participation in the 'community' by contributing comments, videos, views and opinions in discussion forums. Like many Web 2.0 spaces, YouTube is keen to emphasise the notion of community and provides a description of its values and mission. For example, it explains that:

> The community is truly in control on YouTube and they determine what is popular on the site. Discovery is at the heart of YouTube and we continue to create new features

> to help our users discover and share compelling content, for example through featured
> videos, related videos, and guest editors. (YouTube, 2008a)

Here the emphasis is on community control and on reflecting rather than
dictating users' tastes and interests. They go on to say that

> YouTube will always be an open community and we encourage users to send in their
> thoughts and comments about their experiences on the site. YouTube understands that
> each and every user makes the site what it is and welcomes them to get involved to help
> create new features and be a part of new developments on the site. (YouTube, 2008a)

Further YouTube is keen to explain what it will not tolerate but is clear that it is
the 'community' members who need to say what is appropriate and what is not.
In this same section of its site, it repeats again the word 'community' and makes
it clear that safety is the responsibility of all users:

> YouTube is a site dedicated to a community of people who enjoy engaging with each
> other. We don't want content on the site that violates people's privacy or sense of safety.
> By filling out our complaint form to the best of your ability, it will help us assist you
> most effectively with your issue. (YouTube, 2008a)

The site caters for such a diverse audience and such huge numbers, it is easy to
see how material that may seem innocuous to one person may well not be so to
another. Unlike Wikipedia, for example, a site that, as we describe in Chapter 8,
asks for all material to express a 'neutral point of view' (Wikipedia, 2008c),
YouTube invites people to express themselves, to be creative and to 'discover the
quirky and unusual' (YouTube, 2008a).

Although the site can be very useful for educators, its purpose is not to serve
that community; it is multipurpose and offers many things to many people, so
that if it is to be used for education purposes, users need to know how to tread
a path that will feel fruitful and safe. We believe the site is very valuable for the
resources it offers teachers as well as learners; for the opportunities it provides
to share film that learners have created and/or edited; for teachers to upload
pedagogical materials for their pupils, and for practitioners to share ideas with
each other.

Despite the case that YouTube promotes the notion of community and
respect for others, operating a reporting system about offensive material, we
have come across video footage and comments that we think is offensive to
many viewers, or, as we imply above, that teachers would not wish to condone
or introduce into their classrooms. Such material is sometimes racist, offensive

or macabre–such as excerpts from English soccer where one player headbutts another–the many comments that follow this video sometimes degenerate into racist abuse. Astonishingly there are multiple examples of recordings of television presenters vomiting; contestants on shows accidentally being hurt and so on. Of course, these latter examples of unfortunate events being presented as comedic runs in the tradition of much situation comedy humour and may certainly not be offensive to everyone; however, such material being introduced at inappropriate moments into classrooms could be disruptive at least, or offensive at worst. Nevertheless we do believe in the value of YouTube as a resource, we would urge teachers to consider strategies for organizing teaching to minimise such disruptions, to think ahead about how to deal with the inadvertent introduction of this material into their classroom spaces, and also to consider even introducing some examples of this kind of material for critical discussion. For example, as Trier (2007b) suggests, one can explore sensitive issues such as racism by using material from the site.

Experienced teachers will know that such events in classrooms are highly predictable, with or without YouTube, and that students are likely to come across potentially harmful material anyway out of school. We would like to feel that schools could play a role in discussing media with unpleasant messages so that students can learn something about how to deconstruct those messages and how to trace safe paths even in the face of undesirable and profane images and comments. Trier talks about how some material that has been banned from some spaces is accessible on YouTube and gives examples of a few cartoons such as Betty Boop that were banned because of their crude depiction of racial stereotypes in some episodes. Trier suggests showing these and encouraging students to discuss whether they think the cartoons should have been banned and why. Beginning with instances of now dated 'retro' popular culture such as these cartoons, rather than with the more aggressive posturing instances of racism elsewhere, means that pupils can look at racism with more of a distanced perspective than when discussing materials that is more contemporary.

YouTube is a useful and rich resource of some fascinating historic material as others have discussed (Carroll, 2008), and teachers, like Trier (2007a; 2007b), will be able to locate materials that has been difficult to access prior to sites such as YouTube making them available again.

In this chapter we talk about the structure and format of the site, the materials available and how they can be used. We also discuss concerns teachers may have about using the site and ways in which these can be circumvented.

Understanding the structure of YouTube

In comparing YouTube with other Web 2.0 spaces, we notice how the space 'feels' qualitatively different. While viewing the videos, the on-screen layout gives strong messages that this is a public space, rather than a private one. In comparison to blogging software, for example, where users can thoroughly cust-omise spaces to express their identity through the layout of the pages, YouTube is much more limited. Similar to blogs, sites such as Bebo and Myspace allow users to choose from a range of templates, to manipulate the fonts and to add images in a range of different locations. In YouTube, however, users simply upload their video material and it takes its place in a video stream alongside others. Within YouTube, customization is limited: the videos are placed in default positions and set in an inflexible template. Further, even when one is a member and signed into the site, the page will always open to the YouTube homepage, rather unlike Flickr, a blog or Facebook, for example, that all open out at the user's own portal within the space. This simple difference gives YouTube much more of a busy, public, 'marketplace' type of feel. As opposed to the cosy more homely space of a blog, this is a place where one is always aware of the presence of others, where just anyone might be watching and where one would be less likely to find longer debates that may be seen on Flickr, for example (Davies, 2007), or discussions in blog comments (Davies and Merchant, 2007).

Nevertheless members are allowed a small amount of space where they can provide a profile and list their interests; adding to this profile, they can also collate a list of links and screenshots relating to their favourite videos as well as links to outside sites such as MySpace. It is possible to alter the template within a limited frame—changing colour and organisation of the template—but there is certainly no flexibility in terms of adding, say, Flickr images or a site meter. The profile has much less prominence than on blogs, for example, and plays much less of a role than video and commentary in helping present a sense of self on YouTube. YouTube seems then to give participants less control over how members organise the online template; it constrains manipulation of the actual space, setting greater emphasis on the viewing or sharing of video content and on the interactivity around the video material.

YouTube and youth viewing

In a recent series of discussions with fourteen- to seventeen-year-olds from a school in the UK, 27 out of 28 of these pupils reported to Julia that the last website they had looked at was YouTube—22 said that they go online daily and that YouTube was their favourite site, and 26 said that if they were allowed to keep only one item out of a television, a cellphone or their PC, they would choose their PC. One who said he would prefer to keep his cellphone explained it was an iPhone through which he could see YouTube and get online easily. The students claimed that you could watch anything important from the television on YouTube, and that through their preferred social networking site they would find out from their friends what was 'hot' and 'crucial' to see. The referrals took the form of either hyperlinks via instant messaging chat facility or through the use of embedded video in Facebook or MySpace sites. Thus at least for these young people, they watched large numbers of videos that were recommended by others, and far fewer that they found themselves via the site's search facility. For this group of young people, the vital videos of the moment included music videos, sporting moments, funny videos that were becoming 'in-jokes' and anything that may be talked about in their social community online or in school.

In this school, as in many others, YouTube is not accessible through school servers. These activities are strictly 'out-of-school' activities yet vital for students in terms of keeping up with peer group discussions and happenings that are very important to their sense of belonging to a social group and their identity. None of the students involved in these short interviews had uploaded video content to YouTube, only a few had commented on the site. This was because not many of them knew people who did actually upload content and that they would comment on the site 'only if we know the person who uploaded it'. One pupil explained that 'to comment on someone you don't know—well that's a bit weird. Why would you talk to them if you don't know them?'. None of the pupils interviewed had any desire to upload content at the moment—'unless someone showed me how to do good films', said just one. Clearly there are many young people who do upload content to YouTube, but we have found it significant that these users' consumption of YouTube is mainly around facilitating, sustaining and creating social interactivity—they watch and they discuss the content in spaces other than YouTube, such as Facebook, Myspace or Bebo. The fact that they prefer videos on YouTube to programmes and films on television suggests that their viewing is not passive; they are interested in being interactive around the texts they consume. Despite wide media concerns around young people becoming involved with strangers online, for this group at least, online activity on YouTube is about 'thickening existing social relations' (Benkler, 2006:369) and exploring what appeals to them as a group. Through such activities, they develop their friendships in new ways and establish additional nuances for a group identity around shared material, which they can discuss in their own, more private, online spaces away from the gaze of other YouTube viewers.

This kind of activity is one that could easily be cloned, with teachers embedding YouTube texts into new online spaces (such as class blogs, or prompts within wikis) for comments and discussion—albeit teachers' choices may be different from those of their pupils in their leisure time. By discussing materials in online spaces, students can view materials more than once and comment in a considered way, benefiting from a different type of interactivity than in plenary discussions in the classrooms.

Often the links that are passed around travel so fast that they are like viruses. Dawkins (2006) used the term 'meme' as a metaphor to describe the process through which ideas spread and develop through cultures. Memes are analogous to genes, in that for a particular idea to be sustained and passed on, a particular culture must be predisposed to accept it; mutation/adaptation will occur in strong memes in order for it to survive in different cultures. 'Selection favours memes that exploit their cultural environment to their own advantage', explains Dawkins (2006:199). Thinking broadly, Dawkins gives examples of memes as:

> tunes, ideas, catch-phrases, clothes fashions, ways of making pots or of building arches. Just as genes propagate themselves in the gene pool by leaping from body to body via sperms or eggs, so memes propagate themselves in the meme pool by leaping from brain to brain via a process which, in the broad sense, can be called imitation. (Dawkins, 2006:192)

We can see how copying fidelity is aided by the internet because of its ability to replicate information and to make it available across vast geographical, social and cultural areas. On the internet, where connectivity is easy, the cross-fertilisation of ideas is rife and it is often difficult to see where a particular idea began, because of the constant weaving of texts, in and out of each other. Memes can be passed on through hyperlinks, through narratives that have meshed together, through online jokes, or through images passed from screen to screen and copied or altered in some purposeful or accidental way. Lankshear and Knobel have discussed a number of well-known memes (2006a) that behave in this way.

In our observations, as shown, for example, in Chapter 4 on Flickr, individuals seek connectivity; they make rather than break chains, and for the young this is particularly crucial. The internet allows them to thicken their social ties by passing on and adding to online memes. At the time of writing, a popular meme on Facebook is to cast a Harry Potter spell on a Facebook friend. This draws on the current heightened popularity of the book series by J. K. Rowling and coincides with the release of a new Potter sequel. The intertextual linking to a popular film that appeals to adults and to children means that the meme can pass quickly around Facebook—happening by means of one individual leaving a message on someone's Facebook, saying that a spell has been cast on them. One of a selection of icons signals that the spell has been left and the recipient is given the option of continuing the game by casting a spell on someone else. This kind of meme is very common in the Facebook space and helps individuals to demonstrate an online sociability, even when there is no actual conversation happening. As with the last example, participation in the meme requires cultural knowledge gained within the site and beyond it in a range of online and offline spaces. It is clear that the spells are just online play; no one believes that a spell has been cast, but the ritual is playful and is about affirming friendship and knowledge of particular cultural and social happenings.

Jenkins (2006b) writes of convergence culture and the ways in which fans' re-purpose and reconfigure narratives and media texts for their own purposes in a kind of 'textual poaching' (de Certeau, 1984). There are many memes on YouTube that would stand up to scrutiny in explorations of irony, puns and inter-textual references that would engage pupils very well and show them powerful ways in which texts work.

YouTube and youth film production

Despite the group of young people described above who had not made any videos, there are a great many who have become immersed in making YouTube videos. Some have achieved notoriety, such as 'charlieissocoollike' from the UK whose films are recorded by YouTube as having been viewed 2,048,706 times, who has more than 75,000 subscribers and who has been on the Oprah Winfrey show in the US and on breakfast television in the UK talking about his work. On his YouTube channel he now even has a film introducing himself and his work, talking about the YouTube phenomenon and how it has affected his life–such as being paid to continue to make videos and host advertising for YouTube around his videos. His videos include such seeming trivia as 'How to Be English' (charlieissocoollike, 2007a) and the 'Winegum Experiment' (charlieissocoollike, 2007b). The films are witty and clever but cheaply made using one camera and simple editing. He usually talks straight to his webcam, in a straight head-and-shoulders framed shot, with very little background in view. He is evidently filming from his bedroom and it is this domestic home-made aspect that also lends an appeal—this is 'charlieis-socoolike' making an unpretentious film for others to watch from their own computers at home. In this way, it feels very much like peer-to-peer communication, but it is clear the audience is far wider than his own age group and that this kind of film is becoming a cultural phenomenon in its own right—a part of popular culture whose embedding codes and URLs travel across the internet from person to person time and again. Willett (2008) cites Silverstone who comments that 'In play we investigate culture, but we also create it' (Silverstone, 1999).

Like a lot of YouTube material, the videos made by charlieissocoollike are low-tech and have an amateurish aspect that seems very appealing; it is a style that seems to deliberately undermine and turn its back on the high-budget special affects we have come to anticipate all the time from cinema. YouTube thus provides a very different type of film going ever closer to the 'slice of life' feel than even reality tv shows. There is also something spoof-like and irreverent to Charlieissocoolike's work; he seems to spoof even himself, certainly his generation (from his name, to his discussion about his hair and his hat that are 'so cool like'); this ability to spoof, to playfully imitate, as Willett (2008:59, forthcoming) explains, derives from cultural understanding. Thus although some may determine that much of the material on YouTube is inane and banal, there is material that may appear to be so but is often highly reflective, incisive and provides fascinating social commentary that many viewers are seizing upon, recognizing its quality and viewing in their millions. Analysis of such work in school would not only appeal directly to youngsters but also provide the springboard for discussion and analysis, and for video response work. Asking learners to produce their own spoofs requires critical reading of one text and the production of another that highlights the assumptions of the initial text. In some cases, this may require very sophisticated tools and ideas; however, it is possible to see ways in which spoof effects can be achieved more easily as we suggest later in this chapter. Certainly Charliessocoolike's videos would serve as an excellent starting point for video making in the classroom, and pupils would undoubtedly find his work inspiring, focussing as it does on small details of everyday life, such as packets of winegums and making cups of tea! As we describe later, his work has served to inspire at least one family to make their own YouTube videos.

Willett has also explored the work of the young and writes about some of the work on spoofs she has seen (Willett, 2008). For example, she discusses the Bentley Bros. who describe themselves on YouTube as 'a small group of people, that make homemade comedy films with nothing more than a cheap camera and editing software'. The group is comprised of brothers, based in the UK and aged between 14 and 20. The films are scripted, often spoofs, and several are as long as an hour. Like charlieissocoollike, they have a huge fanbase, with over 5,000 subscribers and their channel has been viewed in excess of 2 million times. Their production of 'Resident Evil4' has over 1,500 comments on YouTube and has been viewed 270,000 times. As Willett observes (ibid:63), their work is reaching a much wider audience as a result of being internet distribution, and people beyond their own rural village in the UK seem to understand their humour. The boys, as is the norm on YouTube, are in the position of writers, actors, directors and producers of their work. They present something of themselves, their identities and where they are located; they already had and have developed through their work a media expertise in reading and producing film, as well as in the manipulation of new technologies. They present their friendships through their use of humour and in this way seem to invite others into that group. In presenting themselves as a close network of brothers with a particular type of humour, they invite others to that group, by sharing their playfulness in their films. This is clearly an empowering process and one that is central to the way YouTube works. By drawing on well-known popular culture texts with global distribution, they draw on existing networks and understandings and create new discourses in relation to these. We believe this is important in the context of classrooms, because not only have these boys provided us with materials to analyse and help us deconstruct the films that they have spoofed, but they also show us how YouTube or other video-sharing sites provide us with new ways of dealing with and responding to film.

Willett, looking at spoof videos on YouTube, found that in her sample of UK-only producers the majority were young white men, aged approximately 12–25 (Willett, 2008) and so there is certainly a space here for girls and young women, non-white men and people from other age groups to step into.. Teachers can help to redress this imbalance with appropriate classroom discussion and work.

YouTube and the cool profs

Within the university sector, some tutors are finding YouTube a valuable tool for disseminating information to students, and for students to report on collaborative or individual projects. Some institutions use YouTube to promote their courses, and a number of academics have been using it to disseminate new ideas.

At the University of Sheffield, two mathematics lecturers known online as 'The Catsters'
(2008) post YouTube videos explaining complex mathematical concepts and procedures.

What is fascinating about this material is the steadfastly traditional teaching and presenta-
tion used to represent the mathematical concepts, with a teacher working to a chalkboard and
demonstrating calculations as she speaks. Praise in the comments is effusive, with contribu-
tors asking questions and suggesting additional items that could be covered. The lecturers
are partly inspired to use YouTube in this way so that students of their very large classes are
able to ask questions using pseudonyms—that is, without the shame of showing they do not
understand in front of others. Sometimes they ask students to watch the videos before the
class, just so that they can assume prior knowledge of certain concepts before they begin
teaching in a face-to-face situation. Viewings increase immensely at examination time and
it is clear that students use the videos for revision purposes. Their videos have been watched
by tens of thousands and are much commented upon and even posted into blogs.

Vignette 7—The Catsters

We have not come across schools using the site in this way, for example, to demonstrate tasks for students so that they can watch at home, or to explain concepts so that they can 'listen again', yet we do think that this could be useful—especially if teachers were brave enough to enable feedback so that students could ask questions and comment. However, we feel that by far the greatest benefits could be gained by involving students in searching YouTube for material to evaluate, as well as in producing video for others, or in producing explanatory videos of their own. Such material might also be used for embedding into wikis or blogs for commentaries—and this would mean that the videos could be left as private within YouTube.

How-to videos

As well as the 'how-to' type of video that the Catsters make, there are a great many others within that genre on YouTube—and charlieissocoolike's work includes some from the 'how-to' genre (Charlieissocoolike, 2007a). So popular is this style of video amongst amateur film-makers that one video-sharing space, 'Videojug', is totally devoted to this genre. One of the reasons why this formula is so popular is probably that the format is so simple to reproduce without any media training. The format is now so widespread that we are very familiar with it and the 'direct-to-camera' talk and the one-shot style are easy to imitate and manage even with only one or two people involved. In addition, the structure is very clear and so long as all materials have been gathered before hand and the camera and sound have been set up, one can successfully make such videos alone and with very few props. A further attraction that makes this style useful for the teacher to incorporate into their teaching is that if students make this kind of film, it positions them as experts and encourages them to become very familiar with the knowledge they need for the film. The making of the film is clearly a very important part of the process as well as of allowing peer teaching to take place.

Within the available models of 'how-to' films on YouTube, as well as work from people such as the Catsters, there is an extremely broad range, such as 'How to apply false eyelashes'; 'How to make a bracelet'; 'How to chose a car'; 'How to remove household stains' and so on. Within the genre, there are a great many that imitate the style of television programmes, such as cookery demonstrations, crafting and exercise videos.

We have seen even the very young involved in the making of such videos, for example, the series by the Perklets has caught our attention—partly because the children who present the films are so young: aged six and eight. We describe their work in the vignette below.

> Two children, filmed by their mother using a domestic camcorder, have produced six videos
> and uploaded them to their YouTube channel. The Perklets' Channel (2008) offers various
> 'guides', such as: 'The Perklets' Guide to Baking'; 'Perklets' Guide to a Day Off School' and
> the 'Perklets' Guide to Hula Hooping'. From the first video in this series (hula hooping),
> through to the most recent, 'The Perklets' Guide to being Two'—(about their cousin), we
> see a marked development in their confidence and an understanding of the genre as well as
> a much clearer sense of their audience. In the latest video, the children, speak more loudly
> and fluently, have planned what to say and seem clearer about what the audience needs to
> know. Since they made their first film—which to date has attracted nearly 600 views—they
> have become aware that their videos are being watched by not only those people they know
> but also by those they do not. They have read the comments on their films—made by those
> who they know, and by those who they do not. They have met family and friends who have
> complimented them on their films and been surprised and pleased at the attention. There is
> a sense then that these children, young as they are, are beginning to understand about Web
> 2.0 and its place in affecting their identity offline as well as in giving them a presence online
> amongst those who they have not met.
>
> Vignette 8—The Perklets' 'how-to' videos

We were able to contact the Perklets' mother who explained how the Perklets came to make the films. She described how she had been on a course at work and had been shown one of Charlieissocoollike's videos, 'How to be English' (2007a), which she had found so funny that she had shown it to her children when she got home. They noticed how many comments charlieissocoolllike had received and the children were very excited by this. Their mother explained, 'I suggested to [the children] that we could try making a YouTube film. The hula-hooping one is the first and least sophisticated, filmed in two takes'. In terms of editing, she explains how the children gradually took on a role in editing:

> G is sometimes not keen about bits of the editing, particularly where I include what should be 'outtakes' as he thinks it makes it less professional.... .

> when we first watched the hula hooping film I thought it was hilarious, and G said he hated it. I said, why, it's sooo funny and he said ' it's not supposed to be funny!'. However, other funny bits have been of his making (the baking one in particular had a lot of input from him).

The children's mother handed over greater editing rights to them, passing over at the same time a responsibility for how the film is intended to be read. The Perket's growing sense of audience is confirmed in their mother's next remark,

> The children fully understand what it is to be online. They think it is fantastic! They love it when we go places (family things usually) and people comment on having seen them on YouTube. They help with the planning and the editing (though I have the last say on that!) and come up with ideas.

Safety issues online

Very often issues of safety arise when thinking about film, the internet and children. There are many guidelines available for schools, teachers and parents about keeping children safe (eg., becta, 2008; Kent Local Authority, 2006). Unfortunately, this has often meant that authorities have ended up banning social networking sites across whole districts–and this is an international phenomenon.

We asked the Perklets' mother about this–her children are both of primary school age and, therefore, would be considered vulnerable. She specifically mentioned the inhibiting aspect of school policies:

> Their friends have watched them, as have their teachers and we have all bemoaned the lack of access in the school itself–wouldn't it have been a great 'show and tell' item for them to have to share with their classmates?

She is not casual, however, about safety precautions and has considered these with care:

> I work in the public sector delivering services for children and young people and am very aware of being careful around images and access to children. We have stringent

controls in place around parental consent to images being used, for example, and are forever fending off media enquiries that potentially take advantage of vulnerable young parents/people etc. There were several measures I took to make the Perklets less identifiable. The obvious one is that Perklet isn't their name and their real surname is not common, so that immediately made them less identifiable. Secondly, as I got better at filming/editing, I was able to cut out bits of film that might make them easier to track. For example, in '[The] Guide to Being Big', there were shots that included the front of our house and our car, including the number plate, and so I cut them out. Also, in 'The Guide to Being Two', I have cut out nearly everyone's faces except for those who had agreed it was ok, or in sweeping shots. I also make sure we don't make mention of where we are located. On that broader issue of who knows who's looking at them...well I haven't worried about that really. As long as they aren't at risk it doesn't worry me who sees them—the same applies at all sorts of public places—swimming pools, parks, school playgrounds etc.

The work by the Perklets illustrates how it is possible to participate in Web 2.0 activities and still take a responsible attitude without spoiling what is produced. These experiences show that they have learned a great deal from watching the videos of others, seeing a model they could mimic and produce material of their own. This phenomenon of the creation of amateur video for distribution across local and sometimes global audiences is something that the young are finding very engaging. It is certainly possible, even via the medium of film, to present oneself without needing to betray too much so that it becomes risky. This is giving many youngsters all kinds of opportunities to enjoy a creative process as well to participate in wider social networks. This can be beneficial without the thousands of viewers attracted by a few and is something we would encourage teachers to think about also.

YouTube in the classroom

It is clear that there are a range of ways that teachers can use YouTube in school but there are many barriers to this—not least that in many places the site is inaccessible, banned either by local authorities, as in the UK, the school boards, as in Canada, and other institutions in the US.

Moving through school classroom spaces, talking to teachers and to pupils, and then through into universities and talking to students and lecturers, we have found a multitude of opinions and reports about YouTube. It is a site that houses many contentious, some would say inflammatory, texts, and yet it also provides a free space that could be fruitful for learning. Clearly there is a series of dilemmas to negotiate if teachers are to consider ways of using YouTube in their classrooms. Just as with other types of site and software we have discussed in this book, such as Flickr, blogs and wikis, there are a number of technological skills that are required, but in order to be fully competent or 'literate' in using the site, one needs far more than the basic technological skills of joining the site and setting up an account, of uploading and accessing content, of creating a profile and 'friending people', of making and reading comments. Beyond these basics, users need to know how to read the content critically, be aware of how one represents oneself and remains safe online, understand how to read meanings across a range of texts and how to access meanings not just from within the YouTube space but also by cross-referencing more widely within YouTube as well as beyond. Certainly the skills of critical literacy, along with a knowledge of how to discern the provenance and purposes of messages—where they come from and what they are trying to do—seem to be especially important in negotiating this particular site safely.

We argue that as young people are accessing these sites anyway, we need to give them the critical skills to negotiate the spaces carefully. They will come across texts that are racist, sexist or otherwise socially unacceptable, and they need to be helped to recognise these as such. Sometimes innocuous-seeming material contains messages that are irrational, condemnatory and excluding; it sometimes takes careful, skilled critical reading to discern the motivation behind such damaging texts and we believe that in monitored, sensitive teaching situations, teachers can provide pupils with the tools to read such texts in discriminating ways. Leaving this kind of education to chance is probably fine for some learners, but there are others who may be easily convinced by the sometimes extremist messages that are increasingly being placed online. Distinguishing the authentic from the inauthentic, as well as the racist from the non-racist, is an important skill to acquire—one that can require time and guidance. Furthermore, we believe that pupils can gain a great deal from producing video texts and sharing them and that they can benefit in many ways from distributing their voices beyond the classroom to others located in different spaces. Getting feedback from those who know you from face-to-face, real-world interaction is one thing, but gaining an accolade from those who do not and are never likely to is another.

· 6 ·

GOOD VIBRATIONS:
FROM NAPSTER TO LAST.FM

Tracing the history of music sharing, and particularly the ongoing concerns about copyright, highlights the significance of music as a cultural commodity. Studies in popular culture show how important music can be as a social marker. Musical styles and recording artists are particularly salient cultural forms for children and young people who regularly use musical preference to signal group identity. Loyalties, values and lifestyles often constellate around musical genres, so that within and between social groups and subgroups music can be seen to play a central role in identity performance (Merchant, 2006). Online social networking around music and music sharing shows how such practices are taking on a new significance in the digital age. Music distribution through MySpace and this site's potential for creating and reaching new markets has attracted considerable media attention. But online interaction also enables the development of affinity spaces for niche musical interests. What these new practices mean for the production and consumption of music and how this can provide a model for school-based learning will be addressed in this chapter.

In some ways music and music sharing in Web 2.0 environments raise some very specific issues. These are mostly associated with issues of legality and the ownership of material. This is because the social networks that form around music tend to focus on the work of established commercial artists rather than on user-generated content. Sharing and copying music is, therefore, highly sensitive for those large and influential corporations who own copyright. Of course, these issues are not restricted to the domain of music—all media, and particularly still and moving image, face similar dilemmas. In fact, in a recent initiative based at MIT called YouTomb (youtomb.mit.edu), researchers are involved in monitoring the videos that get removed from the YouTube site because of copyright allegations. Legal wrangles are a common feature of the relationship between music and the internet. In March 2007, Viacom, the company that owns MTV demanded that YouTube remove more than 100,000 unauthorised clips from its site. Although some preliminary agreements between the two companies were made in July 2008, legal discussions are likely to continue for some time.

Despite the legal battle between MTV and YouTube, we are yet to witness such high-profile litigation in the area of video piracy as we have seen in music sharing, epitomised in the case of the Napster controversy. At the heart of these issues lies the 'rip-burn-mix' capability of new media that means that almost any material can be copied and reused without acknowledgement, by almost any computer user with a very basic level of knowledge of the tools. Social networking, which, as we have seen, often depends on the sharing and exchanging of cultural artefacts, puts copyright under the microscope.

Music sharing and copyright

Music sharing has become a particularly sensitive area because pirate copying of music directly attacks the commercial base of recording companies. The issue is further complicated by the fact that illegal copying of music has a long and difficult history—one that includes bootleg vinyl, reel-to-reel tape copying, cassette compilation, and more recently CD burning. As Frost (2007) argues, online music distribution simply cuts out the 'largest claimants to the revenue stream' in a single stroke. Moreover, exchanging digital products not only sidesteps the retailer and the distributor, it also makes the work of producing the physical artefact more or less redundant.

Litigation against the music-sharing forum Napster became a test case in the protection of copyright. When Shawn Fanning and Sean Parker designed the Napster system, which made available the music stored on one computer to users on other computers, no matter where, it seems that they had little idea of the trouble that this would provoke. Essentially, the Napster system built up a database of who had what music. Then, by searching for a particular song, a user could access a list of who had that song and whether or not they were online. Providing the 'holder' of the song was online at the time, Napster could connect the two computers to facilitate an instant download. In this way, peer-to-peer file sharing was made easy—but musicians and particularly their record labels were cut out of the equation. Lessig (2001) documents the legal battle that ensued and the implications for internet copyright.

Copyright laws are supposed to protect property and income and are particularly important in making sure that those who create ideas, designs or works of art are remunerated for their efforts. But with the increasing number of ways that you can copy and disseminate material in new media, these laws do seem to be becoming harder and harder to enforce. Although the Napster enterprise was simply facilitating the exchange of music between peers, the music industry saw it as a system for 'stealing copyrighted material' (Lessig, 2001). In the final denouement of the Napster case, the court ruled that 'Napster was not responsible for contributory infringement unless the copyright holder made Napster aware of the violation' (ibid.). The effect of this was a gradual attrition of the Napster database. Napster, bought out by a German company soon after the court case, is now a subscription service although it does offer some streamed music free of charge as a trial (napster.com). Commenting on the implications of the Napster case, Lessig (2001) suggests that the legal actions brought by large corporations may actually stifle the free and democratic promise of the internet, creating, in its place, a 'more perfectly controlled' environment of surveillance and prosecution.

Although music sharing that uses the Napster principle still continues (for instance, with BitTorrent and LimeWire), it is dogged by issues of litigation. Nevertheless, music sharing is extremely popular, particularly within the 15–25 age range. Survey studies on both sides of the Atlantic bear witness to the vast scale of music sharing. An estimated 60 million Americans, and tens of millions of people outside of the United States, have downloaded songs using file-sharing programs. This in itself raises some important issues for educators. Firstly, in terms of professional awareness of the popular digital practices of children and young people, it is hard to ignore the significance of digital music. Although music sharing does not require particularly sophisticated ICT skills, it is a very good example of the everyday use of digital media and illustrates the current divergence of schooled ICT from the experience and interest of young students. Secondly, as long as there are issues of legality surrounding music sharing, education has a key role in highlighting these and alerting pupils to the issues and possible dangers. If nothing else, school-based work on music sharing can serve as an early and relevant introduction to ideas about copyright and ethics, about the rights of producers and consumers, and about the more adventurous ideology that informs Creative Commons (see creativecommons.org).

Of course, we need to bear in mind that whilst sharing music that is under copyright is illegal, buying it from a provider such as iTunes is perfectly legitimate—as is downloading and sharing music that is not under copyright. Some music is freely available on MySpace, and sites such as Garageband (garageband.com) and Freeplaymusic (freeplaymusic.com) host free music to allow performers to promote themselves. Over the last few years, there have been a number of high-profile news stories about recording artists who, by making available their compositions free online, have built a large enough fan base to attract a recording contract. Similarly, some organizations and artists have taken the decision to embrace file sharing, recognising its rich potential. For example, Berklee School of Music, an organisation that subscribes to the view that file sharing is here to stay, has taken the decision to make their online tuition files free for exchange (Evangelista, 2003).

Music and identity

Popular music is an important topic because if its central role as a marker of identity. Previously the site for adolescent and youth identity performance, music is now significant for all age groups. Music is, as Willis (2000) observes, a potent, global cultural commodity that is used for 'symbolic work' in contemporary society. Popular music builds on a nexus of social practices in which the taste and value associated with kinds of clothing, hairstyles, poster artwork and even body adornment cross-relate (Willis, 2000)—and, of course, social interaction takes place around these social objects. As boyd, succinctly puts it: 'Music is social glue among youth.' (boyd, 2007) and functions as a fluid marker of what is cool and what is not.

From this point of view, popular music consumption illustrates a principle of social networking that we have already observed in our analysis of Web 2.0 practices. What we refer to as a social object, or cultural artefact, comes to play a central role in the affinity space that provides the context for social participation and learning. As new technology becomes more pervasive in the lives of children and young people, the trend for these sorts of practices to migrate to online spaces is likely to continue. Web 2.0 developers have been quick to design online spaces that incorporate some of the practices that are associated with the consumption of popular music. Last.fm—which allows you to 'show off your taste, see what your friends are listening to, hear new music' and so on—has now become a social networking community in its own right (last.fm). A similar service is provided by Pandora, which is based on a music-discovery program that searches its database for similar tracks (pandora.com). In Pandora, listeners submit details of music they like and the service responds by playing similar selections. Users provide feedback on the individual song choices that Pandora 'learns about' and uses in compiling subsequent selections (see below for a fuller discussion).

These developments are indicative of the interrelationship between real and virtual spaces in the performance of identity. It seems important to us that educators are aware of these practices and begin to explore ways of using the social capital that children and young people have developed in educational settings. Dyson's work with young children (Dyson, 2003) provides an insightful commentary on how, under the right conditions, cultural resources such as music and screen narratives can enter classrooms, and writers such as Buckingham (2003) provide a model for this under the umbrella of 'media education'.

Changing patterns of music consumption and distribution

From the point of view of media studies, looking at music in the digital age opens up a rich arena for exploration. Issues about the availability and enjoyment of music—whether about performance rights, the price of concert tickets, or corporate domination of the music industry—provide insights into the ecology of cultural production. Inevitably, these issues are tied up with the technology that continues to play a key role in the history of music performance, recording and distribution. Within recent history, we have seen a massive expansion of radio-based music broadcasting, including the rather troubled emergence of the DAB format, rapid changes in music storage (from vinyl and reel-to-reel tape, to eight-track, audio cassettes, mini disks, CDs and MP3 files), and changes in how live music is produced. All the above have contributed to new ways of consuming music and have altered the role that music plays in our lives. Listening to our favourite tunes on an MP3 player or mobile phone allows us to take our music with us and this compares rather unfavourably with the idea of lugging a carrier-bag full of vinyl LPs around with us to share with our friends.

Despite the emergence of new formats and consumer practices, older values still inform a music industry that continues to fight hard against changing ways of listening to and sharing music. These values have also influenced the ways in which music is packaged, from the design of the CD—to look like a scaled-down vinyl disk—to the way in which music is 'bundled' in the album format. As Frost observes:

> It is difficult at this point to anticipate how consumers will in the future prefer that their music be bundled, if at all. Based on the iTunes and Rhapsody model, we might see the unbundling of the package we think of as an album (a format that itself was sized according to the capacity of 12-inch, 33 1/3 rpm LPs half a century ago) ... . (Frost, 2007:5)

One of the attractions of music file sharing is that the user can select the specific track wanted, rather than the whole bundle, and go on to produce a personalised compilation of his or her own favourites. This was a practice that first emerged with the availability of cheap recordable audio tape and is now reproduced online. Muxtape (muxtape.com) is a Web 2.0 music-sharing site that encourages its members to exchange compilation collections. But alternative patterns of consumption are emerging all the time. The CD multiplayer brought in the idea of randomly shuffling through a bank of CDs, but, as Alderman (2008) observes, the iTunes shuffle function takes randomness to a new level, as listeners are able to draw on their entire music library.

Accompanying these changing music technologies has been the rising popularity of the music video, and the advertising or movie soundtrack. Music video has become a potent way of marketing and remarketing artists, and the increased use of the music soundtrack has harnessed the multimodal potential of film and video to suggest new meanings. In the soundtrack, mood and emotion are often created by the careful use of familiar and unfamiliar music. This can help recording artists to reach new audiences. In a similar way, the association of music with brand products can help the sales of the product and the music.

Web 2.0 and music

Pandora and Last.fm are two popular music sites that incorporate Web 2.0 features and show how the future of music consumption and distribution may develop. As we saw earlier, both these services work on the 'recommendation principle.' So you, as a user, specify a favourite artist and the service delivers streaming internet audio of similar music. In both services, you create an individual profile that outlines your musical taste. You can also, of course, add profile details and so on. In these and related ways, users create a sense of identity or *presence* that is characteristic of many other Web 2.0 spaces.

Although the scope for *modification* is limited in both these sites, the home page that you create builds this sense of presence. For example, the home page of Last.fm displays the last ten tracks listened to, information on each track, name of the artist and time that you listened to it. The service also automatically generates your top ten artists over a weekly period and displays various related information about your online activity. The tagging function, described in Chapter 4, also allows you to label a song (e.g., as *80's New Wave* or *Saturday Night*), and tag searches enable you to look through the 3.5 tracks on the Last.fm database.

There are subtle differences in the ways in which Pandora and Last.fm enact the 'recommendation principle' and these merit some examination here. Pandora uses musical data such as melody, harmony and vocal character to achieve this recommendation—part of what it rather grandly describes as the 'Music Genome Project.' In contrast, Last.fm is a social recommender. It does not collect knowledge about a particular song's musical qualities but depends upon the wisdom of the crowd, assuming that people's musical tastes overlap in useful and productive ways. Although these services do not explicitly depend upon *user-generated content*, the nuances of one's musical taste, the ways in which these are aggregated, and the whole concept of social recommendation are underpinned by the idea of active consumption. So, for instance, in Last.fm, you can access three 'stations' or track compilations—your ordinary radio station, your favourites and your neighbours' station (see below for an explanation of the 'neighbour' concept).

Any *social participation*, beyond simply sharing the information outlined above, is not essential in these music services. Last.fm encourages you to add friends to your profile pages and also identifies 'neighbours' for you. Neighbours are other members who listen to more or less the same kinds of music that you do. This dimension of social networking is extended through joining and creating groups—and again these are user-defined—whether these are about Female Voices, SvenskaPOP, or breakbeats.

Sites such as Pandora and Last.fm are perfectly legal, providing details and links to official downloads, which, of course, users have to pay for. The commercial interest is evident in a number of ways. Last.fm carries advertisements, and the artist-profile pages could be considered as promotional material. Some commentators have even suggested that chart popularity may be susceptible to commercial manipulation, particularly since the CBS takeover of Last.fm in a $280 million deal in 2007. Nevertheless the sites provide a legal forum for music sharing and an interesting template for social networking around media preferences.

On the decks: Between production and consumption

Most of the discussion in this chapter has so far been concerned with the exchange of pre-recorded digital music. The shifting patterns in the way in which we listen to pre-recorded music have also changed the ways in which we think about live music. After all, if we have access to high-quality recorded music, useful and portable storage systems and sophisticated ways of broadening our taste and selection, how does live music fit into the equation? This is a question that lies outside the scope of this book but is certainly worthy of some consideration. So, in this next section, we narrow the focus, concentrating on the role of the DJ—and we do this for several reasons. Firstly, we want to underscore the significance of the DJ in popular music, particularly the playful uses of technology that DJ-ing involves; secondly, we want to review the practice of the DJ as a metaphor for some significant contemporary cultural practices.

The significance of the DJ in popular music performance and consumption is worthy of mention here, not only because successful DJs are arbiters of musical taste, but also because of how their practices are influenced by new technologies and modes of distribution. Since the late 1960s when Jamaican DJs and producers began modifying studio recordings and adding vocal accompaniments (a practice known as 'toasting'), the DJ's ability to recontextualise music has become a revered art. The influence of DJ styles derived from African American oral traditions is also woven into the history of rap and RnB styles of music. Central to contemporary genres of these and other kinds of dance music are the ideas of sampling and remixing.

Sampling, depends upon using or adapting a previously recorded sound or musical extract and incorporating it in a new recording. Typically, this is mixed in with beats and other elements to produce a sort of musical collage. As Dyson explains:

> Sampling is a means of borrowing and manipulating sounds to construct new mixes, new pieces. ... [Rappers] took their samples from previously recorded songs and used them as a background beat for an improvised street poetry. (Dyson, 2003:172)

Often these recordings are in turn remixed live by DJs who may introduce fresh combinations of tracks, add their own samples, or manipulate the tempo, pitch and other musical characteristics of what they play. Although some DJs prefer to work with vinyl, increasingly technology plays an important part in preparation and performance. Although samples are exchanged online, and music technologies, such as softsynths and DJ programmes, play an important role, the significance of DJ culture to Web 2:0 is more in the characteristics of DJ practice. Often described as the 'rip-mix-burn' approach, the remix and mash-up ethos underpins a lot of Web 2.0 development. In this sense, the DJ becomes a metaphor for the Web 2.0 user/developer (Boutelle, 2005).

Lankshear and Knobel (2006a) chart the ways in which remix, as a creative practice, has now extended to cover a variety of forms of cultural production or 'writing'. In an argument that recalls the Bakhtinian notion of doublevoicing (Bakhtin, 1998), they suggest that:

> At the broadest level, then, remix is the general condition of cultures: no remix, no culture. We remix language every time we draw on it, and we remix meanings every time we take an idea or an artefact or a word and integrate it into what we are saying and doing at the time. (Lankshear and Knobel, 2006a:107)

Lankshear and Knobel go on to illustrate how remixing now extends to new combinations of digital image, text, sound and animation using fanfiction, photoshopping and AMV as examples of this. In a more technical sense, the customisation of Web 2.0 applications, the design of widgets and software mash-ups are extensions of this phenomenon.

Issues for educators

Including popular music in education has always been a contentious area and there could be a number of reasons for this. Apart from general professional reservations or prejudices about the value of popular culture, there are more specific concerns about 'colonising' the playful and sometimes irreverent or rebellious themes that emerge in popular music (for a fuller exploration of some of the underlying issues, see Marsh and Millard, 2000). Nevertheless, media education provides some useful approaches, and more resources, such as the Popular Music in Education blog (2008), are becoming available.

Lorraine is Head of Music in a large 11–18 school in an urban area in the UK. She takes full advantage of music-sharing by requiring her students to upload their own music compositions.

She uses numu.org to encourage students to evaluate the work of classmates and other peers using the comments facility. Her students love to listen to their work at home and often share it with family members. Lorraine encourages her students to record and upload their music in the one-hour timetabled music lesson. Students are required to blog their reflections on the composition process as a homework task.

Vignette 9—Music sharing in the classroom and beyond

As we can see in Vignette 9, legal file sharing can be used in the study of music in ways that take full advantage of online activity. But, of course, it is also the case that educators have a responsibility to alert students to the potential risks involved in using and downloading from sharing sites. Becta, the UK government agency that promotes the use of new technology in education, provides useful advice on file sharing (Becta, 2008). Although they suggest that blocking sharing sites in educational establishments is a local decision, they set out some of the potential risks as follows:

- exposure to inappropriate content
- exposure to inappropriate contact
- breach of intellectual property rights
- exposure to viruses and hacking
- high demands on network bandwidth (Becta, 2008).

Amongst the learning approaches suggested by Becta is Websafe Crackerz (websafecrackerz.com/nickstar.aspx), which is a spoof world to help young people make informed decisions about 'using the internet and engaging with people online, common myths and misconceptions about file sharing, risks of viruses, legal consequences of illegal file sharing, and information on spyware and adware.' The site also provides some links to legitimate resources for file sharing.

It seems appropriate to end this chapter on a positive note, which may enable us to understand the power and potential of file sharing. In engaging with the potential risks, it is easy to lose sight of the fact that sharing copyrighted music is only one aspect of file sharing. Applications such as BitTorrent provide a system that allows those who own content to distribute in an efficient way—it is simply a distribution service that distributes the bandwidth cost of large files. And so, by comparing user interests or tastes, they are linked directly with digital products such as text files and programmes. This peer-to-peer linking is suggestive of a new kind of knowledge economy—one in which resources for learning become the social objects around which networks of participation can grow.

· 7 ·

VIRTUAL WORLDS AND
REAL FUTURES

The distinctiveness of virtual worlds as online spaces has led to the development of a body of writing and research that is sometimes quite remote from discussions of Web 2.0. Virtual Worlds are online spaces where participants usually adopt one or more avatars and interact with both objects and other users. These spaces can vary dramatically from each other, with some being total fantasy spaces where anything is possible, while others are very similar to 'real-life' spaces where the avatars' capabilities are similar to those of people in the 'real' physical world. Because virtual worlds share many of the features of videogaming, particularly of Massive Multiplayer Online Gaming (MMOG), attention has tended to focus on their similarities (Carr et al., 2006; Dovey and Kennedy, 2006; Gee, 2007; Steinkuehler, 2008) rather than their differences. An alternative perspective sees virtual worlds as three-dimensional social networks—online environments where users meet and interact with each other and collaboratively create and edit virtual objects. Although some virtual worlds incorporate games, the focus of this chapter is on the characteristics of 2-D and 3-D virtual environments rather than gaming in general. MMOGs and those environments in which predefined roles and rules are built-in are not considered here.

To consider virtual worlds as Web 2.0 spaces, it is perhaps worth revisiting the four characteristics of presence, modification, user-generated content and social participation that we listed in the introductory chapter (p.5) and applying them to virtual worlds such as Second Life (SL) and There. Firstly, in terms of *presence,* it is quite clearly the case that once a resident is logged on, his/her avatar is visible to others in the virtual space (assuming, of course, that the avatar is set to be 'visible' by the user in that session). Through the choice of avatar, one develops an online identity that is, as time goes on, recognisable by others. Over longer stretches of time, this avatar may develop a particular narrative through the history of its activity and interactions and become a character in its own right (Schroeder, 2002). The relationship of this avatar identity to real life, as in other Web 2.0 environments, is quite variable. Some users are motivated to establish a sense of continuity between online and offline life by reproducing some of their real world characteristics in their avatar, whilst others are more interested in creating an alternative identity (Thomas, 2007).

Secondly, some but not all virtual worlds offer a level of user *modification.* Avatar design, as we have described above, is one facet of this. SL provides quite sophisticated tools for designing your avatar, whereas other worlds use a selection of 'off-the-shelf' avatars, as in the case study that follows. In addition to this, some but not all virtual environments are 'mashable', working together with other Web 2.0 applications. So you can, for instance, display Flickr photographs in SL, and the Slurlmarker application is used to bookmark SL locations (Cashmore, 2006). SL commentators tend to see this kind of modification as an important development point in virtual world technology (Bestebreurtje, 2007).

Thirdly, in terms of *user-generated content,* it must first be noted that virtual worlds vary in functionality and, of course, residents too vary in their skill level. So, in some worlds, such as the educational application of Active Worlds described below, the possibility of creating new content for the world is limited. In contrast, the SL environment is built by residents. In fact, it is claimed that, of the 400,000 person-hours logged in SL each day, around 25% are spent creating items for the virtual world (Cocker, 2007). In SL, creating, exchanging, and selling new content is a highly significant activity.

Finally, one of the key features that virtual world play depends upon is *social participation.* Virtual worlds are highly interactive environments that allow residents and visitors to communicate in a number of ways. In most of these environments, this involves synchronous chat (which involves all in the immediate vicinity), whispering through one-to-one synchronous chat, and various forms of in-world messaging (such as one-to-one telegrams). These features assist in promoting social interaction, whether of a frivolous nature or in the interest of more serious collaboration.

From this analysis, we are able to see that virtual worlds have a number of features in common with other Web 2.0 environments and that these features may be exploited to a greater or lesser extent. As three-dimensional social networks, virtual worlds offer exciting possibilities for learning, and these possibilities have begun to attract the attention of educators (see Steinkuehler, 2008; Merchant, 2009). To date, most of this work has focused on older learners—particularly in the HE sector—and in professional training. Work focused on school-aged children is less visible. The Schome Project is one example of this. Schome aimed to document the use of Teen Second Life and to give participants 'a lived experience' of radically different models of education (Schome, 2007). Case study work summarised below and published by Merchant (2009), and Marsh's (2009) study of young children's engagement with virtual worlds, provide some useful starting points.

Getting another life

Current market research shows the growing interest in virtual worlds. At the time of writing, SL statistics claim nearly 15 million residents (Linden Research, 2008), making this the most popular virtual world of the moment. The growth in SL population has been one of the most significant and hotly debated issues in recent internet history. Considering that SL was launched only in 2003, it has made a significant impact on many people's lives and attracts regular press coverage. A virtual economy has developed through buying and selling in-world products or objects; the currency, the Linden, now has an exchange rate on real-world money markets. Many commercial enterprises have opened premises in this virtual world, either to promote their real-world business or, as in some cases, to sell branded virtual goods. In-world advertising has become a phenomenon in its own right. For example, Figure 11 shows how Warner Brothers are promoting Regina Spektor, the New York-based singer/songwriter in SL. In a similar way, pressure groups, political parties and educational establishments have all been keen to have a presence in SL. SL has now also become an online space for academics to meet in and for students to learn in, and a legitimate site for new kinds of research.

Although impressive, the virtual world phenomenon isn't simply about SL. New virtual worlds are continually being developed, and, as the KZero statistics

Figure 11.   Warner Brother advertise Regina Spektor in Second Life.

show (Figure 12), the biggest growth area is the 10–20 age group, with a significant number of new developments in the under-10 age range. A recent estimate suggests that there may be as many as 150 virtual worlds that are targeted at children. Currently the most popular of these are Club Penguin, with over 12 million registered users, and Barbie Girls, with about 11 million (Marsh, 2009).

Although the massive popularity of virtual worlds continues to attract media attention, there is little work that explores their educational potential in the school sector. In higher education, and in academic circles, innovators are beginning to explore learning in virtual worlds, but there is very little work with younger learners. Drawing on a case study of the classroom work of elementary school children involved in a virtual world project, the rest of this chapter will look at the challenges and possibilities of distributed collaboration and problem solving made possible through such work.

The case study that follows is based on classroom use of a 3D virtual world using the Active Worlds software, originally designed as a sophisticated web browser. Schroeder (2002) describes a 3D virtual world as:

> a computer-generated display that allows or compels the user (or users) to have the feeling of being present in an environment other than one they are actually in. (Schroeder, 2002:2)

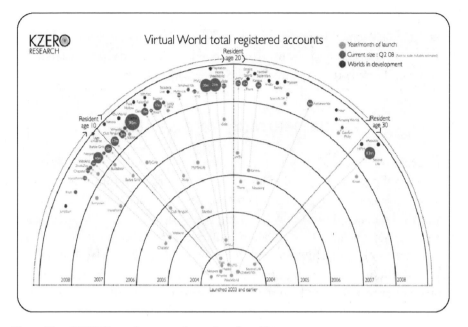

Figure 12. KZERO's market research on virtual worlds.

3D virtual worlds give residents and visitors the impression that they are moving around in an alternative reality, and this immersive experience has, in itself, attracted considerable attention as we shall see. Quite quickly exploration feels less like controlling a figure on a screen and more like being 'in' a world. On some occasions, it can actually feel as if one's avatar has taken on a life of its own. This phenomenon is well documented in the literature (Dovey and Kennedy, 2006; Schroeder, 2002;).

The use of virtual worlds could well enhance or transform learning, yet empirical research that investigates its learning potential in classrooms is still in its infancy. Although there are a number of claims about the high levels of learner engagement in gameplay (Squire, 2002) and the construction of 'powerful learning environments' in virtual worlds (Dede et al., 2006), there is clearly scope for more empirical evidence to back these claims. Some researchers have claimed that immersive environments may lead to distraction and a subsequent loss of focus (Lim, Nonis and Hedberg, 2006), but there is, as yet, insufficient evidence to reach firm conclusions. Early studies such as those of Ingram, Hathorn and Evans (2000) focused on the complexity of virtual world chat, and Fors and Jakobson (2002) investigated the distinction between 'being' in a virtual world as opposed to 'using' a virtual world—but still little rigorous attention has been given to the learning potential of these environments. The work of the Vertex

Project (Bailey and Moar, 2001), which involved primary school children in the UK, makes some interesting observations on avatar gameplay but placed its emphasis on the ICT learning involved in building 3D worlds rather than in the learning and interaction that might take place within them.

Virtual learning—a case study

An educational virtual world project, initiated by a UK local authority in Barnsley, aims to raise boys' literacy by an adventurous and innovative use of new technology that foregrounds digital literacy (Merchant, 2008). In partnership with the company Virtually Learning (virtuallylearning.co.uk), the project team—a group of education consultants and teachers—designed a literacy-rich 3D virtual world that children explore in avatar-based gameplay (Dovey and Kennedy, 2006). The children, in the 9–11 age range, work collaboratively to construct narratives around multiple, ambiguous clues located in the world and, as a result, engage in both on- and off-line literacy activities. The virtual world, called Barnsborough (Figure 13), is a three-dimensional, server-based environment that is explored from multiple but unique perspectives through local Active World browsers. Navigational and communicational tools that are built into the Active World browser enable avatars controlled by the pupils, to move around in virtual spaces such as streets, buildings and parks, to engage in synchronous written conversations, and, in this particular example, to discover clues in order to build their own narratives.

Pupils in 10 different project schools have been using this 3D virtual world, interacting with each other using the Active Worlds' real-time chat facility. The world itself consists of a number of interconnected zones that are life-like and familiar—in fact, they are usually modelled on real world objects. The zones include a town, complete with streets, alleyways, cafes, shops and administrative buildings—some of which can be entered. Outside the town, there is a park with a play area, a bandstand and a boating lake. Visitors can wander through woodland and discover hidden caves. In the suburbs, you can explore a residential area with a variety of housing. Some houses are open for users to investigate and provide further information on the activities and lives of former residents—in fact, the whole history of Barnsborough. In some of the connecting zones, pupils may encounter other sites such as a large cemetery, a medieval castle and a stone circle.

Rich media, tool-tip clues, and hyperlinked and downloadable texts provide clues about previous events in Barnsborough, suggesting a number of reasons

Figure 13.   Screen shot of avatar interaction in Barnsborough.

why the former inhabitants have rather hurriedly abandoned the area. Some possible story lines include a major biohazard, alien abduction, a political or big business disaster or suggest something more mysterious. The planning team has seeded these clues throughout the Barnsborough environment, drawing on popular narratives such as Dr Who, Lost, Quatermass, the Third Man and Big Brother.

In this example, a 3D virtual world provides a stimulating environment for online exploration and interaction. Barnsborough is *designed* as a literacy-rich environment. To enter Barnsborough is to become immersed in a textual universe and to participate in what Steinkuehler (2007) has described as a 'constellation of literacy practices'. Below is a list of the main kinds of digital literacy encountered in the virtual world. These are not directly used for literacy instruction—with the exception of the hyperlinked texts, which are quite deliberately tied to national literacy objectives.

Environmental signs and notices
This material forms part of the texture of the 3D virtual world and is designed to create a real-world feel to the visual environment and also to provide children with clues. Examples of this include graffiti, logos, signs and notices, posters and advertisements.

Tool tips

These give additional explanations or commentaries on in-world artefacts and are revealed when 'moused over' with the cursor. Tool tip messages that draw attention to environmental features ('looks like someone's been here'), hold navigational information ('you'll need a code to get in'), or provide detail ('cake from Trinity's') are shown in text-boxes.

Hyperlinked texts

Mouse-clicking on active links reveals a more extended text. Examples include an oil-drilling proposal (a Word document), a child's diary (a Flash document), and a web page on aliens. Some of these links are multimedia (such as phone messages and music clips) whereas others provide examples of different text types, such as text messages and online chats.

Interactive chat

This is the principle means of avatar interaction and involves synchronous chat between visitors to the world. Comments are displayed in speech bubbles above the avatars heads as well as in scrolling playscript format in the chat window beneath the 3D display.

The Barnsborough virtual world experience foregrounds some important issues relating to engagement with digital literacy in the classroom. The most significant of these dilemmas stem from the fact that it introduces pupils and teachers to new ways of interacting with one another. So, for instance, in-world pupil-pupil interaction is not only conducted in the emerging informal genre of interactive written discourse, but it also disrupts ideas of conventional spelling, turn-taking and on-task collaboration.

New relationships between teachers, pupils from different schools and other adults have also been significant in this work. Issues about authority and what kinds of behaviour are appropriate in a virtual environment have been quick to surface, and this in turn has raised issues for teachers who are understandably concerned about the safety of their pupils as well as how they might monitor children's online experiences and interactions. These online digital practices can, therefore, give rise to uncertainty, particularly where they do not easily fit into established classroom routines. Squire, in an article on the educational value of video gaming, suggests that:

> the educational value of the game-playing comes not from the game itself, but from the creative coupling of educational media with effective pedagogy to engage students in meaningful practices. (Squire, 2002:10)

This observation could apply equally to 3D virtual worlds. In and of themselves, these technologies cannot *create* new forms of learning, but as educators become more familiar with their affordances and the ways in which they are being used in recreational and work contexts, they can begin to experiment with educational uses, to design specific environments and to envision new pedagogies.

Virtual worlds in classrooms

Working in a virtual world brings teachers face-to-face with some complex and challenging issues. In the introductory chapter we explored how new technologies prompt us to rethink some quite fundamental assumptions about educational practice. In our research into virtual worlds, we recognise that it is no easy matter for teachers to abandon the comfort of conventional, classroom-based student-teacher relationships, to experiment with new and fluid online interactions. Virtual worlds can be unfamiliar and chaotic environments in which established routines and control strategies are of little use. Teachers will have to take risks when using these sorts of technology, and that is something that is not always encouraged or supported in our education systems. Furthermore, there can often be technical challenges in virtual world work, particularly in cases in which networks are poorly maintained and connectivity is unreliable. As we have seen in our exploration of other Web 2.0 spaces, local administrators concerned over internet safety may prevent access through the use of filters and firewalls. Vignette 10 illustrates the compromises that one teacher had to make in order to develop some classroom work on virtual worlds.

Mr Jones was interested in introducing virtual world work into his primary classroom in Manchester but was not able to find the financial support to become involved in a major project. Determined to make some progress with his plans, he decided to develop a programme of work that involved his students in discussing and evaluating virtual worlds. He began by demonstrating the Active Worlds educational site (AWEDU) on the class whiteboard, using a visitor avatar and providing a commentary to his class. Students in the class found this exciting and in the discussions that followed, some began to talk about their own use of popular virtual worlds. Mr Jones built on this demonstration by asking his students to discuss what makes a 'good' virtual world and then to refine these into evaluation criteria. They then used the criteria to assess the attractiveness and functionality of a number of these popular virtual worlds. The students visited:

Club Penguin (clubpenguin.com)
Build-a-bear (buildabearville.com)
My Tiny Planets (mytinyplanets.com)

They then made observations about the design of the home pages looking carefully at the visual and verbal messages, examining the 'overt purposes' (what the producers said about the

world: who it was supposed to be for and what it promised) and 'covert purposes' (messages about: who you are, what you should be interested in, do or consume).
The students went on to ask a number of questions in the following areas:

- *What opportunities for interaction/communication are available?*
- *What level of control do you have? (Can you customize/change your avatar or change the environment?)*
- *How varied is the world you explore?*
- *How easy is it to navigate? (How do you know where you are?)*

Vignette 10—Evaluating Virtual Worlds

The experience of the Barnsborough virtual world (Merchant, 2009) also drew attention to the limitations of IT resource allocation in schools. Limited hardware and existing timetable structures made it difficult to achieve the level of immersion and flexible online access required. As others have observed (e.g., Holloway and Valentine, 2002), schools may need to rethink computer hardware location, access and use. In common with other digital literacy practices, virtual world gameplay invites a more flexible approach to curriculum organisation and online access. In cases in which collaborative work between schools is envisaged, additional planning and coordination is also necessary. One of the most important features of digital literacy is its potential to connect learners with others outside the immediate school environment, but we have learnt how this places additional demands on teachers in different locations.

Real and perceived risks may also be faced when engaging in virtual world gameplay. Issues of child protection and the possibility of parental censure received careful attention in the Barnsborough project. Parents were kept informed about the work and the project team was well aware of the moral panic about gameplay and needed to carefully rehearse the educational rationale for the project.

· 8 ·

WIKIS: THE DEATH OF
THE AUTHOR?

A BBC news reporter is interviewing a British member of parliament (MP) in his London office; the MP dresses formally in a shirt and tie and sits grandly at a desk situated in front of bookshelves displaying a range of hefty-looking books—a long row of ancient, leather-bound encyclopaedias. This background library lends the scene a serious air of authority and scholarly expertise—a weighty context for the spoken words of the MP. In this setting, the signification of a series of bound encyclopaedias housing knowledge and facts that have been trusted over generations illustrates the interconnections between print literacy, education and power. These tomes on the shelves even evoke a ritual scene: the process of selecting the correct weighty volume, respectfully drawing it down from the shelf, turning the tissue-thin pages with care, while studiously consulting the text. Such a scene is easy to conjure—associated with learning and a 'respect for books' and their textual authority—it is a well-rehearsed series of actions. The key point about these texts is that they are trusted and that the provenance of the knowledge is assumed to be worthy. The authors are identifiable and are experts in their fields; the information has been checked and verified; the publishing company is well-known and even the very materiality of the volumes—their weight, their size as well as their preciousness—seems to render their actual contents, the knowledge itself, as permanent, lasting and immutable. These are the books often referred to when critiques of web-based information are made. They are the trusted texts, used by generations of scholars and, therefore, proven to be of value. Yet the trouble, as we often find to our disappointment, is that encyclopaedias can become out of date, the facts obsolete (e.g., referring to countries that no longer 'exist') or disproved (e.g., referring to genetic codes as indecipherable) and access to them is often impossible or inconvenient (e.g., their size and weight are prohibitive) or they are too expensive to buy.

Wikipedia—continually updated but never complete

Enter Wikipedia, with its ever-changing bank of knowledge, access that is free of charge to all Internet users and in an 'always on' situation. Wikipedia, probably the best known wiki in the world is the best source of information about itself:

> Wikipedia is written collaboratively by volunteers from all around the world. Since its creation in 2001, Wikipedia has grown rapidly into one of the largest reference Web sites, attracting at least 684 million visitors yearly by 2008. There are more than 75,000 active contributors working on more than 10,000,000 articles in more than 250 languages. As of today, there are 2,528,797 articles in English; every day hundreds of thousands of visitors from around the world make tens of thousands of edits and create thousands of new articles to enhance the knowledge held by the Wikipedia encyclopedia. (Wikipedia, 2008a)

The article 'About Wikipedia', from which this quotation comes, is, of course, an excellent case in point. The statistics can continually be updated so that readers can remain informed about changes almost as they happen. Similarly, science articles can be updated and online references given to other more specialised sites or sources of information. In terms of size (but not materiality), this online encyclopaedia must surely be the biggest in the world; it has the highest number of readers and authors and it is not anywhere near complete. Like many digital texts, it is published—yet incomplete—and has articles that many authors and editors will go back to, change and even delete—for example, unreferenced or highly opinionated pieces will be deleted (Wikipedia, 2008c). Many of the credentials of traditional encyclopaedias seem to be the converse of what is offered online—an untraditional medium with text that can be altered at a moment's notice. It is as if, for some, this makes online text untrustworthy, yet we would argue that what may seem like textual and factual vulnerability to some simply reflects our world that is forever changing. This type of publication constitutes a world in which ordinary people are credited with expertise, and where such expert knowledge can be continually evaluated against the evidence of other texts provided by a wide audience of potential authors/readers distributed across the globe. Everyone can contribute, and everyone (or no one) is an expert. That is to say, irrespective of the identity of the writers, the text undergoes the same kind of scrutiny across the board and is subject to change.

As well as being able to provide information about almost anything one can think of—from aardvarks, to zoology, eBay and memes to Barbie dolls and Johnny Cash— Wikipedia's repertoire is huge and ever expanding. Since it allows immediate publication of events as they occur, Wikipedia is used by some as a way of following global current affairs. Indeed, Wikipedia is known for its ability to provide news coverage around the world, with so-called citizen journalists providing images and 'eyewitness' accounts from scenes of catastrophe, war, and other events, often before 'official' reporters arrive at the scene.

'Citizen journalism' has become a buzzword since the development of Web 2.0. Any individual with access to technology (especially mobile technology) can now instantly publish online. For example, they can cover news of events from where they are at any time and they can publish it to their blog, or to Wikipedia; what is valued is the ability to immediately publish as events are happening, and that journalists are 'ordinary people' who are 'in situ'. Events need not be dramatic, nor disasters, they may be ongoing situations, or local campaigns, but the example we offer here, to show how Wikipedia works, is the posting about the 7th July London bombings (Wikipedia, 2008b).

The case of the 'July 7th London Bombing' article

The facility for immediate and mobile publishing to an online space meant that in the terrorist attacks on London in July 2005, a member of the public at the disaster site was able to take images of the event on his mobile phone (Tagg, 2005) and then transmit them to his blog. These images, first published on a moblog (Alfie, 2005) and later transferred to Wikipedia (2005), were uploaded even before the photographer was rescued from the bombed tunnel. The original blog post along with its many comments can still be read (Alfie, 2005)—and this post transmits a powerful immediacy of the real-life drama even now, over three years later—so that, like many other parts of this entry in Wikipedia, the source can be traced back and cross-referenced against other testimonies and evidence that have been drawn upon. The capacity provided by new technologies to report events as they happen not only has clear practical advantages (helping rescue operations, for example) but also means that information can come directly and 'intact' from those closely involved, rather than being always mediated by professional journalists.

The Wikipedia article about the 7th July bombings was altered many times during the course of the rescue operation and in the weeks following. It was being continually revised as new information became available. Many people were able to contribute to the text, accessing Wikipedia from a range of places. Although the shared and distributed authorship makes the text more comprehensive, editing was later needed to give shape to the various pieces of information, organising the many narratives to give a cohesive account pieced together from evidence gathered and supplied from a range of sources. The text as it stands in August 2008 (as we write) benefits from such periodic reorganisation, clarification and restructuring (including, for example, an entry about arrests made as late as May 2007), and the article now remains one of a series about terrorism in London—and it is more than likely that the article will be amended at later dates. It has been connected, through hyperlinks, to other reports and posts in Wikipedia and is thus now thoroughly embedded into Wikipedia. Further, as mentioned above, it is also linked to a range of sites across the web (and by August 2008, it had accumulated 51 hyperlinked end references and many more in the main body of the text), so that these sources also can be checked, cross-referenced and compared. The existence of hyperlinks seems to encourage readers to trace back the provenance of the entries and to research further, so that the culture of online reading is to travel (or 'surf') from text to text, gathering information, comparing accounts and gaining background data. In this way, critical reading becomes exciting and 'usual'—the hyperlinks show the routes to follow.

As with all Wikipedia articles, the discussions about amendments and additions to specific entries are available for viewing to anyone who registers on the site; a full history of amendments is also archived, so that readers can clearly see disputed information as well as why decisions have been made about particular phrases or words used. For example, there is a discussion whether to refer to the 7th July London bombers as 'suspected bombers' or just 'bombers'. Debate within the discussion area is concise, specific and careful, sometimes as here, at word level, and sometimes concerning the authenticity of various pieces of evidence. The ability to trace the formation of a text in this way is not normally available to readers of other types of published text; the lessons it offers to readers about the process of writing and about criticality are immense here. Participants can, for example, think through the nuances of referring to people as 'bombers' or 'suspected bombers' and what this may say about the writers and their position in relation to the topic.

For many readers of Wikipedia, both the shared authorship and citizen journalism mean that they can hear about world events (and other things) from the point of view of 'ordinary people' and do so immediately. It exemplifies a new kind of text where the ability to publish immediately, even from remote locations, and collaboratively by ordinary people is something new—text that has multiple authorship and one in which participation is valued and readers are encouraged to be writers as well.

Certainly encyclopaedias never attempted to provide a news service to its readers. Yet today's online encyclopaedias (online dictionaries and other such resources) are able to promise an almost simultaneous reporting of events and discoveries, about world records being broken and the like, with which past publication procedures could never compete.

Participation, authorship and anonymity

What Wikipedia gains in terms of immediacy and 'up-to-date-ness', as well as its level of participation, it perhaps loses in other respects. For many readers, their esteem and trust for books is partly founded on the assumption that their authors have been carefully vetted and their work selected, edited and/or reviewed by expert publishers working in the field. They choose texts partly because of what they know about the reputation of particular authors and they can read in a particular way according to what they know about that author. This understanding of the relationship between meanings of books and authors is such an ingrained part of their practice of reading that they find Wikipedia difficult to trust, for example, it is not possible in our writing here about the 'bombings' article to reference specific authors of the Wikipedia entry—we have cited it all as '(Wikipedia, 2008a)'. The elusive identities of the entry's authors mean for some readers that the quality of the work cannot be easily assessed.

Certainly not being able to establish a specific author goes against the academic tradition of clear citation and the educational emphasis on individual contribution. Such a tradition of providing an author name, a date and even a publisher name and place and other details for all citations is a required practice for academic writers of peer-reviewed journals and books. This is a tradition where credibility of knowledge derives partly from the proven scholarliness of the writer and his/her ability to cite reputable evidence and to set out original ideas within the context of other research in the field. For some, the consequence of a move away from this convention is, therefore, that they cannot trust what they read. Further it is not clear how to identify the date of publication, since the text itself is in a constant state of flux and publication is theoretically never complete. The text, as has been stated, began on 7th July and has been altered many times since. Nevertheless, we argue that Web 2.0, in providing us with the opportunity to produce texts with shared authorship and with distributed authorship (across time and space), gives us texts that have a different kind of value than the kinds of texts that have the sort of (usually) single, accredited authorship we are accustomed to. With just a small amount of expertise, it *is* possible to look 'behind' a wiki text and to not only contribute but to also trace the history of the discussions about composition as well as see even the names of contributors to such discussions. These names are hyperlinked so that interested readers can click through to contributors' profiles, follow links to other sites (such as contributors' blogs) and see other Wikipedia entries they have participated in writing. The use of this kind of Web 2.0 profiling provides readers with a way of looking at authors' credentials that may be new to traditional publishing but is becoming ever familiar as a way of authenticating identity in online social networking, of building trust and establishing an online identity/reputation.

Bias and neutrality

Unlike editors of newspapers, or indeed of books and academic articles, the team behind Wikipedia asks its contributors to write in a 'neutral' way and to try not to push a particular viewpoint or to persuade readers to take on a particular opinion. It explains specifically:

> Neutral point of view is a fundamental Wikimedia principle and a cornerstone of Wikipedia. All Wikipedia articles and other encyclopedic content must be written from a neutral point of view (NPOV), representing fairly, and as far as possible without bias, all significant views that have been published by reliable sources. This is non-negotiable and expected of all articles, and of all article editors. For guidance on how to make an article conform to the neutral point of view, see the NPOV tutorial; for examples and explanations that illustrate key aspects of this policy, see Wikipedia: Neutral point of view/FAQ. (Wikipedia, 2008c)

Many of the terms used within this rubric are hyperlinked with further explanation and examples such as 'Wikimedia', 'reliable sources' and 'NPOV tutorial'. Neutrality is, of course, hugely problematic and is certainly a difficult objective to achieve. Many contemporary academic writers have dispensed with such an approach to their writing, recognizing that neutrality is impossible and thus embracing an approach that argues in favour of a 'partisan' demeanour, saying that as long as the author's viewpoint is clear and does not make a pretence at objectivity, this is writing at its best—honest and responsible (Carr, 2000; Greenbank, 2003).

In writing about educational research, for example, Carr proclaims:

> Far from being some kind of unwelcome intruder whose presence or absence can be empirically detected, partisanship is an essential ingredient in educational research whose elimination could only be achieved by eliminating the entire research enterprise itself. (Carr, 2000:441)

Carr's premise is that we cannot look at anything or report on anything in a value-free way, and, simply put, to claim to do so would be dishonest. Presumably, Carr (who writes about Educational Research) would critique Wikipedia's desire to claim for non-partisanship and that an 'NPOV' is possible. Wikipedia further explains that:

> Wikipedia: Neutral point of view is one of Wikipedia's three core content policies. The other two are Wikipedia: Verifiability and Wikipedia: No original research. Jointly, these policies determine the type and quality of material that is acceptable in Wikipedia articles. Because the policies are complementary, they should not be interpreted in isolation from one another, and editors should familiarize themselves with all three. The principles upon which these policies are based cannot be superseded by other policies or guidelines, or by editors' consensus. Their policy pages may be edited only to improve the application and explanation of the principles. (Wikipedia, 2008c)

Thus unlike most other pages, the principles described here are immutable and incontestable. Despite our difficulty in accepting the feasibility of achieving an 'NPOV' in any writing, we recognise the value of what Wikipedia is aiming to do, which is to provide a text that aims to make its authors continually accountable and reflective. Wikipedia warns on every editing template that 'If you don't want your writing to be edited mercilessly or redistributed for profit by others, do not submit it,' and that 'Content that violates any copyright will be deleted. Encyclopedic content must be verifiable. You irrevocably agree to release your contributions under the terms of the copyleft free licence'. Certainly shared authorship requires many compromises to be made but there remain many other spaces where polemic and strongly worded points of view are welcomed. Wikipedia is simply not this kind of space.

Wikipedia is not without its critics, and some sites have been set up in opposition to it (Conservapedia, 2008a) or at least to parody it (Uncyclopedia, 2008) and so it is not surprising that the most ardent of these may attempt to harm its integrity. Conservapedia critiques Wikipedia, for example, because of its alleged anti-American, anti-homophobic and pro-atheist (for example) stances, and that 'despite its official "neutrality policy", Wikipedia has a strong liberal bias' (Conservapedia, 2008b). As examples, Conservapedia cites the following:

> Wikipedia's entry on Barack Obama claims that he 'was selected as an editor of the law review based on his grades and a writing competition,' [25] when in fact the *Harvard Law Review* has long used racial quotas for admission.

...and that....

> Wikipedia has an extensive entry on 'Creation myth'. [42] Describing Creationism as a 'myth' is yet another attempt to disparage Christians, and although the theory of evolution satisfies Wikipedia's definition of 'myth', Wikipedia never describes it as a 'myth'.

Such critiques of Wikipedia, as offered by Conservapedia, provide interesting areas for investigation by students who might be invited to discuss the difficulties associated with attempts to write with an 'NPOV' and how even when densely edited, text will nevertheless find it hard to escape some kind of ideological trace. It would be interesting, for example, to compare entries across the Wikipedia and Conservapedia and discuss similarities and differences.

In this chapter, we argue that the way in which wikis are authored can plant a healthy suspicion of all authors; that is to say, they may help us realise the need to read all texts critically. We argue that learners in the classroom be encouraged not just to read wikis but to also be involved in contributing to them—learning a sense of responsibility about text making as well as understanding how important it is to try and find a variety of sources of information in order to assess knowledge and evaluate what is being said. In realizing the culpability of individual wiki authors and the vulnerability of all information, readers should hopefully be led to realise that all authors are equally culpable and that all texts should be read, perhaps not always with suspicion, but certainly with a critical eye.

Vandalism

Despite Wikipedia's best intentions, because of its easy accessibility, the site is frequently vandalised. The fact that anyone can contribute is democratic but also problematic. Individuals sometimes purposefully change text so that it is wrong, move pages about by altering the links, connect words to inappropriate content or photoshop images. Wikipedia has a reasonably lengthy post devoted to defining vandalism on its site and provides directions to users about how to respond to vandalism they notice (Wikipedia, 2008d). We have certainly come across anecdotes, perhaps urban legends, where individuals have been persuaded of particular facts by being directed to a Wikipedia entry that had been created specifically to persuade a particular individual of a particular fact. Some of these may be totally fun 'practical jokes', or could clearly have dangerous repercussions. Like all public spaces, whether digital or not, wikis are open to abuse and it is up to the public to monitor and report on harmful activity. Educators have acknowledged for a long time their role in helping the young become responsible citizens, and online nettiquette and responsibility now need to become a part of this. Scrutiny of and involvement in wikis can help work in this area and, as Richardson argues (2006:61), 'there are vastly more editors who want to make it right than make it wrong'.

We are aware of the vast amount of criticism that Wikipedia receives in many education circles but we have seen Wikipedia being used in classes where students were encouraged to improve upon existing entries after careful research. This has been very successful in schools where 'wiki-sceptics' have complained about the accuracy of certain entries and then shown how they can as 'responsible citizens' improve upon offending inaccuracies. As Richardson (2006:64) comments, 'If your student writes a great paper on global warming, why shouldn't she add what she found to the global warming entry on Wikipedia?'

Wikis—so what?

Deducing from what we have shown about one specific wiki—Wikipedia—the reader may now be able to draw up a quick checklist of what a wiki actually is. We would itemise the following:

- The text can be edited by anyone who is registered on the site.
- Individuals who set up the site can set out specific rubric, guidelines and community values for others to follow.

- Authorship is shared and distributed.
- Editing discussions and histories can be archived and consulted.
- Openness is valued.
- Collaboration is valued and individualism is less valued.
- Wikis are in a constant state of flux.
- Text can easily incorporate links to other sites, to entries on its own site and to profiles of contributors.
- Referencing is highly valued.
- Incorporation of texts and items from other sites is endorsed—as long as legally adopted and sources are cited.

As Lamb (2004) explains, there are arguments about what true 'wikiness' is, but it is possible to come up with principles such as those we have listed above. The derivation of the term wiki is variously explained as being an acronym for 'What I know is' and as the Hawaiian word for 'quick' and so does not necessarily help with definitions. Lamb observes that:

> As wiki usage grows in popularity with other online cultures, … scores of emerging wiki systems are adding functionalities such as restricted access, private workspaces, hierarchical organization, WYSIWYG (what you see is what you get) Web editing. (Lamb, 2004:3)

This development in easy-to-use wiki software has meant that there is a broad range of types for teachers to choose from. They can set up wikis so that they remain private to specific groups, or so that although public to read, only specific people can edit them. The essential feature of this wiki software, which allows multiple users to co-create interlinked pages, provides a rich resource through which students in geographically dispersed locations can learn about each other and collaborate on shared interests (as Vignette 11 illustrates).

Jeroen Clemens of the Helen Parkhurst School in Almere, in the Netherlands wanted to investigate the learning potential of Web 2.0 software and so he set up a partnership with a fellow enthusiast Anica Petkoska, a teacher at the Gostivar Secondary School, in Macedonia. They developed the MacNed Project using free wiki software provided by wikispaces.com. The wiki involved students in building a shared understanding of their own patterns of engagement with new media that they wrote about on the wiki. The wiki, written entirely in English, allowed both groups of students to develop their language and communication skills. They demonstrated and shared ICT skills by embedding videos and other media. Through this work they were also able to build intercultural understanding by encouraging students to write about their own lives, interests and experiences.

Vignette 11—Wiki sans frontieres

The MacNed Project illustrates how the new ways of communicating and collaborating that characterise Web 2.0 can be used to develop learning. Although it could be argued that the same kinds of understanding could be developed through more traditional approaches, the possibility of co-constructing text in different geographical locations, exchanging and commenting on work in different media, creates a heightened sense of interactivity and a more overtly participatory space for learners. The work also begins to point to a changing role for educators who, in this case, needed to coordinate the work and provide the context for interaction—in short, to design the learning experience and encourage participation.

We have also seen other examples of wikis that have been set up by teachers, to which pupils contribute over time. For example, an Australian primary school teacher set up a private wiki for her class and by way of introducing the new project writes on her class blog:

> Our Wiki's (sic) is different to our blog. Our blog records news and stories of interest to our classroom so that others can read about it. The wiki is a place to record lots of information that you might want to add to over time. You will all be able to edit the wiki but only people we invite will be allowed to look at our wiki.
>
> An example of a wiki is Wikipedia. Wikipedia is like an encyclopedia but anyone can edit it. If you disagree with something someone has written in Wikipedia you can discuss your opinions on the discussion page. If changes are made they are recorded in the history page. Here is a link to a video that explains how Wiki's work.
>
> Here is a link to the our How to Wiki wikipage.
> The types of information we will collect on our wiki will include:
>
> - Technology tips
> - Useful links
> - Instructions on how to do things at our school (Wormbins, 2008).

This teacher uses a broad repertoire of Web 2.0 tools in her classroom and organises and communicates about them through her blog space. The wiki is one that is intended to support other learning activities that the children are involved in—and in involving them in a wiki, the pupils act as researchers supporting their own learning. This is enquiry-based learning that places the children in control. They are positioned as experts and as advisors; yet despite the authority and responsibilities they are given, the collaborative nature of the work means that they can support each other by editing each other's work and seeing the project as 'ongoing' in a perpetual state of revision.

Wikis can be kept from one year to the next with new groups of students alternately using the existing text and then also editing and adding as time progresses. This is the way the Geography project worked in Vignette 12.

Lynne Reynolds set up a wiki for her class as part of a project about the locality of her school. The school was in a tourist area of rural England but had suffered loss of trade in recent years due to so many people preferring to go abroad for the good weather. Many of the pupils' parents were in the tourist industry and they were aware of the importance of the place looking good in order to attract visitors. They had strong (but differing) views about how the place should be cared for in order to maintain a decent standard of living for the local population.

The teacher set up terms that she wanted the students to define and explore but also later allowed them to add their own terms. She began with general geographic terms such as 'local', 'urban', 'environment', 'tourism' and 'industry, but the pupils felt the need to create links within their work to definitions and explorations of terms such as 'litter', 'pollution' and even 'regeneration'. As the project continued, the wiki became quite rhizomatic so that there was a whole section that developed around the history of road names. This all became relevant as the pupils started to realise that contemporary problems and solutions (such as local flooding) could have been predicted by noting that streets and avenues had names such as 'Springwell Grove' and 'Brook View'. Photographs and videos were embedded into the wiki illustrating aspects of 'pollution', 'vandalism', 'regeneration' and so on.

The pupils added a link to Google maps discovering the usefulness in the detail of the 'Satellite' view. Pupils took photos of the area in their own time and were able to post them to a Flickr stream, just in case they would be useful to the project later.

Parents and others in the local community were given access to the site and they volunteered to be interviewed about their views. The site was maintained over the whole academic year and as a starting point for pupils working on a new project in the year following.

Vignette 12—Local Geography

The format of wikis is so flexible that they can be used to set up to meet very local needs and interests, such as 'Our School'. Different authors will have different ways in which they can contribute—from parents and pupils, to school cooks/janitors/caretakers and ancillary staff. Teachers can use wikis to help students with exam preparation. One local teacher we met, who was on maternity leave during the pre-exam period, encouraged her students to build a wiki for microbiology revision. In this way, she was able to view and comment on her students' revision work from home.

Wikis can also be used, to structure works of fiction that *do* become complete as in the vignette below. Here a young teacher, keen to introduce Web 2.0 into his classroom, wanted to develop participation through a narrative-writing project.

This work was successful in introducing young children to the tools provided by pbwiki.com and motivated them to write in new ways that his young students found particularly motivating. Although this project did not aim to develop much dispersed work (the text was composed in the classroom), it was shared with a wider audience in ways that were greatly enhanced by the technology.

Andy Cave, a newly qualified teacher, started working with 8-year-old children on the outskirts of a large town in the North of England. He was keen to develop new uses of technology. Building on existing plans in literacy, he helped the students to understand the defining features of science-fiction story writing by looking at film extracts and stories. Andy then developed a collaborative writing project that was wiki-based (using pbwiki.com). His students worked in small groups for this task. The groups worked in turn to compose chapters, each of which was a new linked page on the wiki. Each group had to read the previous chapter and then write the next. Other groups used the comments feature to give feedback, to provide new ideas and to point out inconsistencies.

The complete story was then edited and illustrations were added. Public access to the wiki meant that students could share their work with parents and extended family, who were also encouraged to leave comments but could not edit the text. Since some of the students had family in different parts of the world, the wiki became a useful way of sharing experience.

Vignette 13—Wiki writing project

Authors' comments

Certainly, as authors of this print text, we remain committed to the value of books and see them as having a major role in the sharing of ideas, knowledge and discussion. However, we are happy to see similar material in a range of formats and understand that different media shape our messages differently. We are clear also that the meanings of texts and the role they can play change when they are mediated through different modes, and we are also aware that different types of text provide different affordances and constraints. The dynamism of wikis and the opportunities they offer for collaboration, for shared and distributed authorship, are something to be prized. Books provide opportunities for strong and clear opinions to be professed and for perusal through words that will not change over time; they are extremely valuable for other reasons as well. As teachers and researchers of literacy, we are keen to promote the use of the whole range of text production and consumption activities and see worth in introducing students of all ages to wikis, partly as they are valuable as a way of sharing information and collaborating over text making. This process can widen students' repertoires as readers and writers, also showing them the value of editing and refining. Further involvement in wikis can help learners become more aware, critical readers, understanding that texts are constructions with particular viewpoints. Teaching learners how to check and evaluate sources within a wiki can of course also lead them to become more sceptical readers of paper-based texts.

Finally we are keen to emphasise the pedagogical value of collaboration over text making because of the way it disrupts the notion that individualised learning and text production is the best way of being a scholar. We are interested in this book in exploring the notion of 'social participation'. Wikis are an excellent example of social participation and what we have come to think of as online interactivity. In the world of Web 2.0, wikis have not enjoyed the widespread popularity that other applications have. The exception is, of course, Wikipedia and this particular site has attracted public attention and plenty of debate in academic circles.

· 9 ·

RESPONSES AND RESPONSIBILITY

This chapter draws together some of the common themes that have emerged in our exploration of Web 2.0 services and explores how they may be able to enhance or transform learning and teaching. We look at this in terms of response and responsibility. Here, response refers to the ways in which educators may be able to harness the power of these emerging technologies to develop new kinds of learning communities—online spaces that reach far beyond the walls of the classroom, connecting learners in different social settings, and crossing the boundaries of age and culture in new kinds of participation. Responsibility refers to the ways in which teachers will need to be aware of how new media also serve the needs of what Buckingham calls 'capitalism's relentless search for new markets' (2003:203), as well as other undesirable forces in society that may ultimately put students at risk.

We begin with a general look at responses, before moving on to some more specific detail. In doing this, we recognize that it is possible to identify different levels of engagement with Web 2.0 in the classroom. These different levels of engagement may well reflect our confidence and competence with new technology. In our work with teachers in a variety of settings, we have become sensitive to the fact that some of the practices that we are enthusiastic about constitute a 'big ask' for some colleagues. As researchers, we have the advantage of having been immersed in some of these Web 2.0 environments, whereas colleagues, often locked into the intense face-to-face business of classroom life, do not always share our experience. For classroom practitioners, the most common initial reaction is to seek for ways of using new technologies within the context of existing practices. Although we suggest a more radical vision, it may well be the case that some colleagues need to build on small successes in their progress towards a more ambitious vision.

The responses outlined in some of the earlier chapters as well as in some of the vignettes of classroom practice we have offered could be broadly grouped into three different categories. These are:

1. *Online activity that replicates traditional print literacy practices.* This describes Web 2.0 work that is unadventurous but safe. Perhaps this sort of response is a starting point. The main difficulty is that it is limited, and students, particularly if they engage in more sophisticated practices out of school, may quickly switch off.

2. *Web 2.0 as a text to study, to explore and to think critically about.* This values the text produced, injects an important critical perspective, but tends to turn vibrant practices into objects of study.

3. *Harnessing the power of Web 2.0 for educational purposes either in Virtual Learning Environments (VLEs—such as Moodle or Blackboard) or on the internet.* This sort of work can be challenging and teachers may face obstacles and even opposition. Edgy and sometimes uncomfortable, this approach opens up new vistas and can be extremely motivating for students and teachers.

These categories roughly correspond to the analysis advanced by Burnett et al. (2006). The first of these sees new literacy as enriching existing practices; the second as extending the reach of critical literacy practices; and the third is concerned with transforming practice.

Using Web 2.0

We have suggested that some of the features of Web 2.0 such as establishing a presence, modifying a personal page, creating and responding to user-generated content, and using the communicative tools for social participation are important ingredients for a new kind of learning. Our descriptions of Web 2.0 practices in this book repeatedly refer to the nature of the socially situated informal learning that takes place. Taking Web 2.0 to school may involve a more focused look at how the affordances of different kinds of services relate to educational objectives. Here are some examples of what we mean.

We noted in Chapter 3 that one of the defining features of a blog is its sequential and chronological format. This suggests that blogs can be effectively used in classrooms to reflect chronology—for instance, in the development of a narrative, the design and implementation of a project, the progress of research, and observations or reflections on processes that develop over time. Blogs, as multimodal texts, allow us to represent this chronology in a variety of media and provide a record that is accessible to a wide audience. Educational blogs can invite participation as they capture quite specific kinds of learning and reflection as they unfold over time. We suggest that blogging really takes off when it involves a range of readers and writers in different settings. There is a growing interest in educational blogging and this suggests that the blog format may have multiple uses in educational contexts.

In our discussion of photo sharing, Flickr in particular, we made quite extensive reference to the kinds of learning that can develop. Not only were we keen to point out the significance of learning about the visual medium itself, we also drew attention to other opportunities that Flickr presents to educators. Amongst these we noted that quite fundamental operations such as labelling, grouping and sorting images were significant in their own right and could also be applied to a wide variety of educational projects and age groups. The way in which written language is used alongside the images and sets is a powerful reminder of the multimodal dimension of onscreen text making. Like blogging, photo sharing offers interactive opportunities when the images are made accessible to others beyond the classroom—we also observed how the meanings of images can change dramatically through the application of different titles, tags and descriptions.

Video and music sharing raise some quite complex issues for teachers— particularly those related to unregulated and copyrighted content. Nevertheless we have been keen to point out how services such as YouTube can be used, and to underline the importance of how learning *about* participation and learning *from* participation can be developed in these settings. In contrast, some rather different issues arise with the use of virtual worlds and these are more to do with their availability in school contexts. Teachers are, for good reason, uncomfortable about allowing free exploration of the more public virtual worlds; therefore, purpose-built environments such as the one described in Chapter 7 are likely to be more attractive. Clearly virtual worlds present rich opportunities for learning, but at this point in time there are financial implications for this sort of project work.

In our consideration of wikis, we observed that this technology provides an ideal environment for developing learning in a wide range of different settings. Social participation and collaborative learning are built into wiki design. Furthermore, the increased availability of wikis that are free of charge to educational institutions means that there are few obstacles to development in this area.

The conditions for learning through participation

Developing authentic Web 2.0 practice in educational settings requires a sensitivity to some of the key characteristics of insiders' uses of these new technologies (Lankshear and Knobel, 2006a). Of course, this idea underpins the work in this book, and we have been keen throughout to show what 'mature' practice looks like, in order to tease out its essential characteristics. Simply signing students up to a service is not enough! We suggest that Web 2.0 work is likely to flourish if careful consideration is given to each of the four quadrants in the diagram below (Figure 14).

We will deal with each quadrant in turn. First, by *purpose* we refer to the need to be clear both about the sort of purposes that Web 2.0 services normally serve and about the specific goals that educators have in mind. For example, there is little point in using a sophisticated application such as a wiki to reproduce repetitive or traditional print literacy practices. So at the very least we would see that an underlying purpose of wiki writing is to develop a knowledge

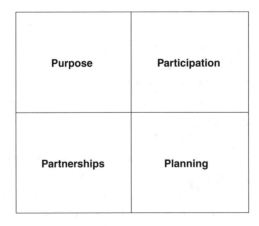

Figure 14. Conditions for successful Web 2.0 work.

resource through collaborative working and some sort of distributed activity. But this, on its own, is not sufficient, because a wiki project also needs a focus or educational purpose. It needs to be about something, and that something needs to be authentic and relevant to students' learning.

Second we need to consider *participation*. Throughout this book we have argued for the use of Web 2.0 spaces as an arena for learning through social participation. As we have seen, the sites and services we have explored all promote participation in some form or another. Learning through social participation may well require a more authentic conception of what constitutes worthwhile knowledge than that found in many curriculum documents (McCormick and Murphy, 2008), but, however the domain of learning is specified, educators can be encouraged to think creatively about different kinds of participation. In some of the vignettes included in earlier chapters, we have seen how students of different age groups and in different locations and cultural contexts can participate in an extended learning community. There is also the scope for the involvement of other teachers, parents and other adults.

Our consideration of participation brings us quite naturally to the idea of forging new *partnerships*. Ever since the pedagogy of writing began to recognise the significance of audience, teachers have been asking themselves the question 'Who can my students write for?' Unfortunately, more often than not, if students are not simply writing for the teacher, they are asked to write for an imagined audience. Despite this, creative teachers are able to find new audiences for students' writing—this is documented in some inspiring work on writing (e.g., Dyson, 2003)—and uses of new technology (Bigum, 2000). Participation goes one step further than this, because it demands more than an audience—it requires co-producers—participants who are able to comment give feedback and generate their own content. In order to achieve this, some of the most successful Web 2.0 projects have invested time in identifying partners, whether they are located in other institutions, in the community or in the world of work.

Finally, no successful educational enterprise can proceed without careful planning, and Web 2.0 work is no exception. The vignettes of practice we have included have all been the result of careful planning—planning that allows for creativity and takes into account students' existing knowledge and experience. And this planning has also been sensitive to the particular strengths of Web 2.0 applications and their adaptability to classroom uses whilst taking into account learning sequences and the kind of scheduling that is necessary to make participation work.

Responsibilities

The word 'responsibilities' may sound a little old-fashioned when applied to the free-flowing world of emerging technologies, but we wish to argue that because teachers have professional responsibilities for their students, they now have the added responsibility of helping them to navigate the online world. Understandably, teachers' concerns are for a safe and orderly space in which learning can take place. Yet, at the same time, the classroom world is a world in which this sense of order has traditionally been mediated by a set of print literacy practices and instructional routines, powerfully structured by curriculum discourse. Disturbing this fragile ecology is a risky business, and strong support and sensitive professional development are required if we are to move beyond some of the curriculum constructs and pedagogical conventions that narrow our vision of learning. Teachers need not be the docile operatives of an outdated, centralised curriculum. As some of the work described in this book suggests, they can be innovative in responding to the potential of powerful new technologies.

Innovation, when it is concerned with Web 2.0 technologies, may mean reconsidering roles and relationships. A shift in emphasis towards participatory learning and shared problem solving requires a different approach from teachers. Furthermore, in an environment in which the learning community extends beyond the classroom, the teacher's position of authority needs rethinking. For example, in a virtual world in which some of the participants are not in the same classroom and not necessarily familiar in a face-to-face context, teachers are obliged to consider how they interact and the role that they then play (see Merchant, 2009). Reproducing traditional classroom roles will not work well and most teachers would not wish their role to be reduced to policing children's online interactions or indeed to marking their work on blogs or wikis.

Even in quite straightforward online work, teachers have to look for alternative ways of designing their students' learning experiences. This is illustrated in some of the student-teacher interviews reported by Burnett (2009). For instance, one of her interviewees, Holly, describes an online session that involved students in a straightforward internet search activity set by the teacher. First we hear from Holly; this is followed by Burnett's commentary.

> We gave them different websites to go to but because they had so many different topics it's quite hard and they ended up just looking at music websites and stuff and you'd constantly be checking every single person to check what they were doing cos you couldn't spend your time with one person. You had to be constantly checking everyone else. 'Look—you can't go on that website- you have to be working'.

Holly suggested that the constraining boundaries of the classroom were pushed as children moved into the more fluid spaces of the Internet and her authority as teacher was challenged as children started exploring their own paths and browsed in a way that did not support task completion. These networked practices however seemed to challenge the teacher identity she aspired to as the children, through their own practices, placed themselves in new power relationships with Holly. (Burnett, 2008:65)

Davies and Pahl (2007) describe having worked with teachers who did feel able to allow students to explore the internet during lesson times in more fluid, unrestricted ways. The learners had been described as 'disengaged' and disenchanted with more traditional curriculum and pedagogy. The teachers in the post-16 setting experimented with using a range of popular culture modes and media and found that they were able to create what has been termed a 'third space' through which the official curriculum was negotiated. The teacher's responsibility to guide pupils' learning, to manage groups and to keep them on task are areas that are, as yet, underexplored in the literature on new technology and, as our examples show, certainly do warrant careful exploration.

The concept of teacher responsibility is also strongly implicated over concerns about internet safety. We do not wish to underplay this area or add to growing moral panics about the dangers of online activity, and so we have dealt with this issue at various points in the book. At this point, it is sufficient to say that plenty of guidance is now available and the vast majority of teachers are well aware of how they might deal with issues of exposure to inappropriate material and child protection in online environments.

A final aspect of professional responsibility is concerned with helping students to develop a critical understanding of some of the values and interests that lie behind the design of Web 2.0 services. This may require introducing some analytical tools of the sort suggested by Buckingham (2003) and is suggestive of the practices described as critical digital literacy (Merchant, 2007). If it is true that one can critique a given discourse only from the point of view of an alternative discourse, then one of the aims of Web 2.0 work in schools should be to provide an alternative perspective on popular blogging, music sharing and so on. However, we would also want to be wary of a crusading approach in which education is seen as a way of 'cleaning up the world' or as a critical practice to inoculate against the presumed evils of consumerism. We would rather see the teachers' responsibility as one of asking questions and guiding learners to a closer examination of what may be taken for granted in the emerging technologies.

Revisiting the rationale for Web 2.0 in schools

In what follows, we revisit the claims and counterclaims set out in Chapter 1 in the light of our exploration of Web 2.0 spaces in order to indicate a way forward. We pick out some of the key themes that Web 2.0 work addresses.

- **The digital divide.** Since Tapscott (1998) first coined the phrase 'digital divide' to describe differential access to new technology, the concept has accumulated meanings. It now no longer is just about differential access to new technologies but also describes attitudes and feelings towards them (see Selwyn, 2004). Whether these differences relate to family background, social class or gender, educators using new technology are well placed to address issues of equity. The key point is that all learners should have a critical awareness of Web 2.0 and, as a result, be in a position to make informed decisions about the nature and extent of their online participation.

- **Repetitive practice.** Some critics have observed that children and young people become locked into a safe zone of repetitive practice in online environments (Livingstone and Bober, 2004). However, these repetitive practices do represent learners' cultural capital and are important resources to build on. There are clear parallels in children's book reading. Young readers may often get 'stuck' in reading and re-reading a narrow range of books; the reading teacher's role is to build on this confidence in the familiar and develop a wider reading repertoire. In a similar way, educators will want to extend students' online repertoire.

- **Preoccupation with the frivolous.** In an investigation of teenage chatrooms, Merchant (2001) discussed whether there was worth in the seemingly frivolous exchanges between teenagers engaged in synchronous interaction in chatrooms. Here, we have argued that educators need to understand and capitalise on the new ways of being and interacting made possible through Web 2.0, and that they also need to investigate the educational potential of social networking. In order to do this, there is a need to conceptualise the difference between casual and frivolous online interaction and those kinds of communication that have the characteristics of 'learning conversations'. Although there has been considerable development in our knowledge about the characteristics of learning conversations in face-to-face interaction in classrooms (Mercer,

2000; Alexander, 2006), there is little equivalent work in the field of online social networking.

- **Attitude to new technology.** Working in the field of new technology often engages us in conversations with enthusiasts, but it is important to recognise that a significant number of educators have had negative experiences of technology either in their personal or in their professional lives—and, of course, the same is true for some pupils. Although confident users of new technology tend to be content to explore and learn in a playful, experimental fashion, others are more cautious. Professional development that aims to introduce specific applications of new technology needs to be sensitive to this, providing plenty of time, support and reassurance to those who are less confident. A survey of teachers' ICT skills and knowledge needs commissioned by Scottish government identified similar themes. The survey found that teachers

  > want good quality training, with hands-on experience, plenty of help and guidance, and opportunities to work with, and share ideas with, other colleagues...[and]...feel they need opportunities, time and ongoing support if they are to gain maximum benefit from training. (Williams et al., 1998)

- **Literacy standards.** Much of the moral panic around new media focuses on the idea that they distract the attention of children and young people from engaging with print literacy practices and are a causal factor in falling standards in literacy in schools and amongst school leavers. Although there is no empirical evidence to support this view, headline claims such as 'Texting and emailing fog your brain like cannabis' (*Daily Mail*, 2005) and scare stories that students are using inappropriate language in national tests because of their online literacy lives still have currency. A more balanced view would be to acknowledge that literacy practices, as social phenomena, are subject to change. The rapid advance of screen-based technology as a site for new kinds of writing has accelerated the rate of change and as a result destabilised commonly held views about the nature of reading and writing. Despite the fact that students' performance in literacy continues to be measured against standards of conventional print-based practices, there is little evidence of the negative influence of new technology. In fact, some research suggests that students now read and write more than previous generations when they inhabit text-rich online worlds. Simply recognising that many

Web 2.0 practices are dependent on written communication is one way of justifying their inclusion in classroom life.

- **The risk factor.** Concerns for the well-being of children and young people are central to the values and role of a professional teacher and it is highly appropriate that these concerns should extend to the virtual environments of Web 2.0 and social networking. However, educating children and young people in internet safety is likely to be far more effective if real experience is provided rather than the alternative of applying blocks, filters and other controls—or even avoiding online activity altogether on account of its perceived danger. After all, we do not discourage swimming because of the danger of drowning; rather we alert children to danger and provide them with the knowledge, skills and behaviours that will enable them to experience the benefits.

- **Social control.** Enthusiasm about new Web 2.0 and social networking environments may sometimes cause us to be somewhat uncritical in our response. It is important to remember that many providers have commercial interests at heart—even in those cases in which technical skill is being shared and distributed free of charge (as in the distribution of open-source software or freeware), it is still constrained by the design and informed by the values of the producers. If users are being unwittingly exposed to advertising, or if market researchers are tracking their preferences, these and other hidden interests need to be explored with students. In this way, educators can introduce an important dimension of criticality into young people's consumption and engagement with new media.

Back to the present

Although some current trends in the school sector are promising, there remain concerns about the slow take-up of new technology (Becta, 2008). This is particularly the case with Web 2.0 applications. Some of the learning platforms (or VLEs) currently being developed in UK schools do incorporate at least some Web 2.0 features—what is lost in terms of broader participation may be compensated for by easy access and the guaranteed safety of such provision. In higher education, robust IT infrastructures have led to the large-scale adoption of learning platforms, and this has brought both advantages and disadvantages. A common practice is to use a VLE as an extension of hierarchical control, either as a platform for institutional or instructional material or, where student contribution is invited, as a tool for monitoring and sometimes assessing student participation in terms defined by academic staff (Burnett, 2008). However, VLE development in environments such as WebCT and Blackboard has led to the incorporation of some Web 2.0 tools. So it is now possible for students to create personalised portals that incorporate feeds from their favourite sites and for staff ('administrators' or 'instructors') to use blogs and wikis for online learning.

We have argued that digital literacies are central to new kinds of social practice and can be incorporated into classroom settings. We have also shown how literacy continues to play a central role in social participation and knowledge building, and how digital connection allows this to happen in ever more fluid and distributed ways. The question of whether the new communicative spaces described can provide an arena for the more systematic and structured interactions that are associated with formal education is not an easy one to answer. After all, classrooms are quite distinctive social contexts in which patterns of interaction and the availability of communicative tools are often restricted or carefully controlled (Kerawalla and Crook, 2002), and so, adopting and adapting digital literacies easily disrupts traditional classroom practices in ways that are unsettling to teachers. Indeed, as Carrington (2008) suggests, alternative learning designs and pedagogies may be required, and these may be achieved only through more far-reaching school reform.

New literacy practices in the classroom contrast starkly with the educational routines of book-based literacy, as well as with the dominant ICT pedagogies that often support the maintenance of centralised control through teacher-led use of whiteboards, instructional software, and highly structured VLEs. Collaborative, peer-to-peer interactions, including communication with those not physically present in the classroom, suggest a very different set of resources and educational concerns. Everyday uses of new technology, particularly recent Web 2.0 developments, raise new questions about digital literacy and its role in education. For instance, what should we teach children about the kinds of online communication that are helpful to relationships and helpful to learning; how can teachers support and encourage peer-to-peer interaction without stifling it, and above all, how can we help pupils to become critical readers and writers in online environments?

· 1 0 ·

CONCLUSIONS

Our intention in writing this book has been to explore the potential of new kinds of online activity for transforming educational practice. In order to do this, we have used the label 'Web 2.0' as a convenient way of capturing new and accessible ways of using the internet, drawing on examples of popular Web 2.0 services to illustrate the kinds of learning they can support. By doing so, we have also made the assumption that 'transforming' education is at the same time both desirable and possible through the use of these new technologies. In this concluding chapter, we turn the critical lens on these notions in order to open new possibilities for exploration and debate.

In much writing about Web 2.0, there are some taken-for-granted assumptions about its distinctiveness and, more often than not, unqualified claims that it is 'a good thing'. In the Preface and Introduction, we were keen to point out that Web 2.0 has become a useful way of referring to characteristics of online life that have been evolving for some time. Indeed, if we were to follow the Berners-Lee critique (2006), it could be argued that the 'social web' was there, at least, in its conception, from the very beginnings of the internet. From this point of view, polarising Web 1.0 and Web 2.0 becomes increasingly problematic (see

Lankshear and Knobel, 2006a). Perhaps a more helpful perspective is one that charts the evolution of a number of different characteristics or uses. Whether or not it would be possible to argue that these can combine in force to reach something like a 'tipping point' (Gladwell, 2000) may safely remain the province of future internet historians.

Characteristics of Web 2.0—revisited

To develop our ideas in this book, we identified some basic characteristics that seemed to us to be useful in trying to capture what we felt was most important about Web 2.0. We recognised their provisionality from the outset and tried to suggest that they might be seen as continual rather than essential criteria that a service needed to meet in order to qualify as being Web 2.0. These four characteristic features that we set out in Chapter 1 have been used in our description of some of the popular sites we have written about. In working with them, we have become aware of two issues. One is the interrelatedness of the four characteristics and the other, quite predictably, is the relative degree to which they are articulated in the sites we have looked at. By way of summary, Figure 15 provides an overview of these observations and notes the roles that these features play in some popular sites that are commonly held up as examples of Web 2.0.

We believe that this shows that these are useful descriptive characteristics for looking at the current state of Web 2.0. However, in researching and writing the book, we gradually accumulated a longer list of features that are worthy of at least a brief mention here and do, in themselves, draw attention to some emerging trends.

A number of these developments suggest ways in which life online and life offline interact and influence one another. For example, in Chapter 6 we noted the way in which random music-shuffling has become a popular way of consuming music. When we looked at randomness in other online spaces, we were surprised to find a number of variations on the same theme. The 'interestingness' algorithm in Flickr, explained in Chapter 4, performs a similar function, as does the 'next blog' feature in Blogger that randomly directs you to someone else's blog. In a way, the 'I'm feeling Lucky' search button in Google replicates this behaviour, too. The random or chance encounter with others or with other artefacts is, of course, an important feature of offline life, too. 'Bumping into' a friend or being introduced to someone at a party works in a similar way. Similarly

	Presence	Modification	User-Generated Content	Participation
Blogger	Profile, posting and feeds, blogroll, comments	Customisation of template, widgets	Posting, links	Comments, RSS feeds, links
Flickr	Profile, 'from your contacts', updates on photostream, your comments, use 'notes', the look/ layout of the photostream	Page layout, address, use of Flickr in other spaces (e.g., Blogger, VoiceThread)	The photostream, titles, descriptions and tags, notes on images, testimonials on your profile by contacts, comments and discussion, Flickr mail	Comments, tags, notes, testimonials, Flickr mail. Forum discussion— community activities
YouTube	Profile, personal channel, updates and comments, the look/ template of your 'channel'	Use of YouTube in other spaces (e.g., in Flickr)	Your videos, tags, comments— verbal and via video response, YouTube messages	Comments, tags. Forum discussion— community activities
Last.fm	Profile	Change the colour of the home page, import 'chart' widget (e.g., to Blogger)	Your playlists and stations, your tags, comments and discussion	Comments, tags, other people's charts
Second Life	Your avatar and its look, what your avatar says and does	Design and modification of avatar	Building, mash-ups, machinima	Interaction with other avatars
PBWiki	Members can track changes, read feeds, comment, add tags	Customisation of template, widgets	The text itself— rich media and writing	Comments, tags

Figure 15. The 4 characteristics of Web 2.0 at work.

just 'picking up' a snippet from the radio or a magazine in a waiting room enables us to chance upon new information or to make creative connections between different pieces of information. The random elements in Web 2.0 applications replicate this experience, mirroring how we stumble upon things in everyday life. Perhaps it is not surprising then, to find an application called StumbleUpon (stumbleupon.com) that does just this with your search engine. So StumbleUpon publicises itself as a:

> downloadable toolbar that beds into your browser and gives you the chance to surf
> through thousands of excellent pages that have been stumbled upon by other web-users.
> (StumbleUpon, 2008)

What is interesting about StumbleUpon is that it incorporates the social element in its operation—and this has been another recurrent theme in our explorations.

Peer recommendation, as typified by allconsuming.net, has been commented on elsewhere, for example, in our consideration of music sharing (Chapter 6); however, with the proliferation of Web 2.0 applications, the challenge to new start-ups is not only to encourage this sort of interaction but also to find ways of populating their sites in the first place. Various approaches that depend upon early adopters recruiting others from their existing social networks have begun to emerge. So, for example, Flickr has a number of incentives for members who introduce friends to their photo-sharing site including discounts for buying Flickr accounts as presents. Facebook, and other sites such as linkedIn.com, use the more intrusive approach of an automatically generated email, whereas the music service Last.fm allows you to send a recommended track to your friend. In some ways, these and other recruitment strategies imitate real-world word-of-mouth marketing and introduction services.

Related to this, as we have observed in passing in earlier chapters, in-group membership and friendship are signalled in a variety of ways in Web 2.0 spaces. The ubiquitous practice of 'friending'—adding someone to your friends or contacts list—has become commonplace. Social ties are thickened by a number of online feedback mechanisms based on commenting, ranking, rating and favouriting (as we saw in Chapters 4 and 5).

Finally, there are the various strategies that are used to keep others informed of online activity. If your friend is regularly visiting your Web 2.0 space, it is relatively easy to keep up to date and to interact. But most people find this very time consuming, particularly if their online life is distributed across a wide range of sites. RSS feeds have proved to be a particularly useful feature in keeping track of multiple sites as they update. Feeds provide you with shortened 'headline' information from sites (blogs, wikis, etc.) that you are watching. So rather than having to regularly check a newsblog, for example, you can keep track of its feeds. This can be done in a number of ways. Guy uses netvibes (netvibes.com), a free service that allows you to create a personal page, or a portal, which you customise to include feeds from favourite sites. In this way, rather than having to visit Julia's blog to see if she's posted anything, he can watch for a new feed and respond if appropriate.

Keeping up to date is achieved in other ways, too. Your Flickr home page shows new photographs from your contacts more or less as soon as they have uploaded them and in a similar way to the familiar MSN 'nudge', Facebook has developed the idea of a 'poke'. The social affordances of Web 2.0 technologies together help to maintain the sort of dynamism that any community needs—a dynamism that is dependent on new events, new interests and new contacts. This must be balanced out with a certain stability or familiarity in which users gain the sense that they are 'going to' a place they recognise, in order to engage in an activity that gives them a sense of satisfaction. In most of the virtual spaces we have explored in this book, members are able to create their own area, to build a home/home page that provides them with a sense of territory. This personal space functions also as a way of establishing a presence and as a place to make public displays of one's interests and affiliations.

Web 2.0 as a 'good thing'

The trends and tendencies in the evolution of the social web—here referred to as Web 2.0—have certainly captured the attention of some academics, commentators and internet enthusiasts, and many of these have made quite startling claims about how technology transforms and will continue to transform our lives. Most of these claims are based on the premise that Web 2.0 and new technology in general are a good thing, creating new opportunities for social interaction and participation, and new possibilities for giving ordinary individuals voice in a world dominated by conservative political institutions and multinational corporations. This emancipatory discourse surfaces in the writing of Rheingold:

> Online social networks are human activities that ride on technical communications infrastructures of wires and chips. When social communication via the Internet became widespread, people formed support groups and coalition groups online. The new social forms of the last decade of the twentieth century grew from the Internet's capability for many-to-many social communication. The new social forms of the early twentieth century will greatly enhance the power of social networks. (Rheingold, 2002:xviii)

And:

> The liberating news about virtual communities is that you don't have to be a professional writer, artist, or television journalist in order to express yourself to others. Everyone can be a publisher or a broadcaster now. Many-to-many communication media have proved to be popular and democratic. (Rheingold, 2002:121)

Although elsewhere Rheingold is cautious about over-optimism, there is a sense in his writing that technology has the power to transform society, rather than the more realistic view that transformation is about changes in what *people do*.

We believe that an unqualified acceptance of the Web 2.0 phenomenon is too simple a response. On the one hand, it would seem to be more reasonable to look at Web 2.0 simply in terms of the emergence of new channels of communication and new spaces for interaction. This implies that services that depend upon user-generated content and social participation are essentially neutral, and value can be judged only by looking at the purposes they serve and the content that is then created. On the other hand, it is important to recognise the influence of producers; we can all benefit by paying careful attention to the values and assumptions that are built into the design of new web spaces. Whether that design concerns the details we are asked or permitted to include on our personal profile, the sorts of cultural artefacts that are deemed to be worthwhile, or the kinds of interaction that are allowed, belief systems and power structures are never too far away, even though they may be well hidden.

These observations provide a clear direction for the work of educators, who need to recognise the growing influence of Web 2.0 in the media lives of the children and young people they teach. Education can draw on students' experiences of new media, it can harness its power to promote learning, but it also has a responsibility to develop a critical understanding of the forces that shape, direct and control cultural practices.

Some observations on social networking

One of the common characteristics of the web spaces we have explored is the various opportunities they provide for interaction between users. The more successful Web 2.0 environments seem to have a built-in purpose around which affinities are constructed. Social networking does not just happen in an unstructured way, it seems to be at its most successful when it is loosely organised around cultural artefacts or what Engestrom (2007) refers to as 'social objects'. In the case of Flickr, the social object is the digital photograph; for Amazon it is the book, and for YouTube, the video. These social objects are the focus of user-generated content and the resultant interaction that takes place. As objects become of particular interest to individuals, a social network develops around them. This is not particularly different to the formation of traditional interest groups—apart from in two aspects. First, because the interaction is online, social networks are often dispersed (time and location are no obstacles to communication), and second, because social networking sites allow for varying degrees of engagement, they lend themselves to lightweight engagement and multiple group membership (for a fuller discussion of the implications of this phenomenon, see Benkler, 2006).

It seems to us that this idea of social participation and networking around a social object holds the key to an understanding of what Web 2.0 has to offer education. As the example of the padlocks (in Chapter 4) illustrates, if we conceive of the social object as a 'learning object', distributed communication and knowledge building can be supported by the tools that constitute Web 2.0. Social participation, as we have described here, forms the basis of a learning community that reaches well beyond the walls of the classroom and involves learners at different times and in very different places. In the final analysis, whether or not this vision suggests an educational transformation is for the reader to decide.

GLOSSARY

Aggregator: Sometimes also referred to as a feed aggregator or RSS aggregator. This is a piece of software that collects feeds from various sources and displays it in a window on your computer. If you want to keep a check on the activity in a number of websites, this is the software to go for. (See also Feed). Using an aggregator means that you can easily access information across many websites in less time than it takes to visit each one. The popular web-portal www.netvibes. com is a quick and easy way of aggregating feeds. It can combine information such as news and weather forecasts with feeds from your favourite blogs. Thus, for example, Mrs Costello has feeds for local weather updates, BBC news, her students' blogs, her Spanish sister-in-law's Facebook and a link to the website of a tribute band 'The 'Counterfeit Beatles'. This means that Mrs Costello has a 'one-stop-shop' online space that she can go to, check whether anyone has recently updated their site, without needing to visit each individual one.

AMV: AMV is a popular form of mash-up based on micro-clips from Japanese Anime (animation) that are spliced together and synched with a music soundtrack. AMV (Anime Music Video) is exchanged and evaluated in peer networks such as animemusicvideos.org. Examples of AMV can also be found on YouTube.

Silvermoon377's video 'She's Just Oblivious' was, to quote Silvermoon, made 'in August 2006, It took about 18 hours during one week'. In the spirit of Web 2.0, she invites viewers to 'download the higher-quality version' on her own website. In making the AMV, she uses film footage from 'The Melancholy of Haruhi Suzumiya' and music from Skye Sweetnam, 'Sharada'.

API: An API is the technical term used to describe the programming that enables one application to work with another. The API (Application Programming Interface) can be likened to the handshake between two different programmes. So, for example, the Flickr API allows you to use your Flickr photostream in other web environments and, of course, to create Flickr mashups. Subscribers to Facebook can, for example, embed content from YouTube or Flickr and even use music applications as part of the template. Facebook users will also often receive invitations to participate in the use of new applications (such as Scrabble) often developed by the Facebook community.

Asynchronous Communication: This term describes a way of communicating that assumes that participants are not online at the same time. For example, on email or discussion boards, messages can be left and responses given at a later time. Thus, for example, Dr Joolz has left a message in the Nag's Head—located in the community section of the eBay site—about a second-hand charm bracelet she has and is asking if anyone knows good sellers of silver charms. She will check back during the week to see if anyone replies.

Avatar: A computer user's on-screen representation of him/herself. This may simply appear as an icon, a photograph or a cartoon figure. In virtual worlds, avatars are often animated figures (such as people or animals) that can move through the world and interact with others. Individuals may have a single avatar that they use in different environments or multiple avatars. When working in the Virtual Space of Active Worlds, Guy has an avatar who wears wrap-around shades, a red T shirt, stone-washed jeans and sneakers (he thinks this makes him look super-cool!).

Blog: A contraction of the term 'web blog'. Kept online, a blog can be private, public or open to a group of invited individuals. Regular postings are organised in date order. Blogs can be kept by individuals or groups; the content can include written text, audio files, images and embedded video. It is usually possible for others to leave comments, but this facility can be disabled or limited to selected users. Blogs can be easily created using free template-based software such as that provided by www.blogger.com. For example, Sabine Chorley, a wedding photographer from Ontario uses Blogger to present some of her professional photography work. She imports images of professional assignment work to promote her business as well as to enable her clients and their families and friends to access images on the web from wherever they are.

Browser: A browser or a 'web browser' is the programme that allows you to access, view and move between web pages. Examples of popular web browsers are Firefox, Internet Explorer, Safari, Netscape, Yahoo and Google Chrome. Some environments such as Active Worlds or Last.fm depend on the use of a specific downloadable browser for accessing their servers. It is possible to customise some browsers so that, for example, Dr Joolz's Firefox has tabs that enable her to move quickly from Flickr to her blog; she also has Google embedded in the toolbar of her browser.

Citizen Journalism: Journalism or documentary by 'ordinary people' who are not professional journalists. Citizen Journalism has been enabled by the development of the internet, especially since the emergence of Web 2.0 software. Ordinary people have taken advantage of the facility to self-publish and to make local events and issues more widely—even globally—known by posting breaking news. For some this represents a democratisation of news publication and production; others see it as the erosion of 'proper' and professional reporting. The writer Jeff Goldstein uses his blog Protein Wisdom to give his personal take on the world and national news. Other types of citizen journalism deal with more parochial matters, such as Nigel Tyrell's 'Love Clean Streets / Love Lewisham' blog that regularly presents posts about the Lewisham area of London and his perceptions of 'problems' on the streets.

Feed: A feed is sometimes also called an RSS feed or RSS. RSS stands for 'Really Simple Syndication'. A feed is a piece of code on a website that others can copy or 'collect', thus enabling them to monitor what is happening on the site. For example, someone may wish to monitor activity on ten different blogs. The codes can be collected and placed on an aggregator (see above) so that a reader can go just to that one space in order to monitor what has been happening on those ten different blogs.

File Sharing: File sharing involves exchanging files between personal computers. This usually takes place in peer-to-peer networks and is most commonly associated with the exchange of music and video files. BitTorrent is a popular peer-to-peer programme used to enable file-sharing between computers. Gezz is an enthusiast for 'acid house' music and regularly uses file sharing to update his personal collection to download tunes to share with his friends. They regularly visit his site in order to check out new tunes and to recommend others, while he draws on a range of sources elsewhere online in order to accumulate a comprehensive collection. In this way, Gezz has established an online reputation as a connoisseur of acid house.

Folksonomy: Related to the term 'taxonomy', this term describes the way in which participants in a Web 2.0 space have assigned tags or labels to the content. These tags identify the prevalent themes, topics or areas of interest for individuals in that particular environment. Aggregating these tags creates a folksonomy. Visitors to the site can then search 'by tag' and see all the objects labelled by a specific tag. Popular tags are often represented in a tag cloud (see below). For example, looking at the tag cloud on Flickr will give an indication of those most frequently used tags. (See also tag and tag cloud).

HTML: HTML is a popular scripting language used in writing web pages and helps in defining page layout, fonts and graphic elements. HTML (or Hyper Text Mark-up Language) is, nevertheless, quite complex and time-consuming—often seen as the domain of specialists. Many Web 2.0 sites allow you to use a simple editing window with familiar word-processing functions, rather than HTML, to write content. Some users (such as bloggers) may learn some bits of HTML to customise their page or to add additional links. So, for example, Debbie put <a href="http://www.flickr.com/">Flickr </a> in her school wiki to make it easy for her class to link to Flickr in order to access images for their project.

Instant Messaging: Shorthand for synchronous online communication. Instant messaging is used to refer to communication between users who are online at the same time. The result is commonly in the form of an interactive written conversation between two or more people. Instant messaging is used in MSN and some virtual worlds. More sophisticated messaging allows participants to embed icons and to share images, video and audio files. (See Synchronous communication). Every evening after school, best friends Lily and Vera communicate about what has happened during the day; although they have been at school together all day, they go their separate ways to their homes in different suburbs of Taipei. Using MSN, they chat about the day's events and exchange opinions and views and are even able to see each other as they talk.

Interoperability: A generic term used to describe the capability of two kinds of technology to work together. APIs (see above) can enable different software applications to work together, whereas mash-ups depend on successful interoperability.

Learning Platform: A catch-all term for online learning environments designed for the education market. These are usually closed or controlled intranet systems. Alternative designations are Learning Management Systems or Virtual Learning Environments (VLEs, see below). Some Learning Platforms also include student-tracking and assessment data—sometimes these integrated systems are called Managed Learning Environments. Blackboard and WebCT are popular commercial VLE tools, whereas Moodle provides free Open Source Software (see below) for building VLEs. In a collaboration between academics at Sheffield Hallam University and Umea University in Sweden, colleagues use a bespoke Moodle environment to exchange documents and rich media texts about their ongoing research.

Mash-up: The term mash-up comes from the hip-hop music practice of mixing two or more songs. In online spaces, a mash-up is commonly found on sites such as YouTube where a video might have mix footage from a number of films and a single backing soundtrack. These are often comedic and frequently become memes. AMV (see above) is a specialised form of mash-up. Another commonly used term for mash-up is 'remix'.

Meme: A contagious idea that is passed on socially; analogous to a gene, a meme will survive if it finds a culture that it is suited to. Dawkins invented this term in *The Selfish Gene* and gave examples of fashion, catchphrases, tunes, popular ideas and so on. These may change and adapt according to the interests and predisposition of the culture that they are successively passed on to. There are many online memes partly because it is so easy to pass on and replicate ideas—as well as to adapt them slightly, for example, through Photoshop and movie-maker programmes. For example, a hugely popular meme amongst bloggers has led to the development of a collaborative blog that collates examples of 'Fail's, such as misspellings of words on notices and labels, misunderstandings of words and visual jokes. Items can take the form of images and YouTube videos—these can be seen at http://failblog.org/

Open Source Software: Software that is made available online for users to modify and develop. Open source projects are often developed by a community of volunteers. However, there are commercial vendors that enhance open source software and charge a fee—the most notable example being Linux. More often than not, open source software is free. Examples are the Firefox web browser (see above) and Moodle (see Learning Platform).

Podcast: This is also known as an audiocast, but the term 'podcast' has become familiar because of the popularity of Apple's iPods that are often used to replay audiocasts. A podcast or an audiocast is a multimedia broadcast hosted on a website. It can be audio- or video-based and is delivered in a format that can be downloaded onto desktop computers or portable devices such as MP3 players and played at any time. Peter James regularly downloads BBC Radio's programme *In Our Time* to his computer. He transfers this onto his iPod and listens during his walk to his workplace; this means he does not miss his favourite programme that is broadcast at a time when he is at work.

Portal: A generic term used to describe a gateway to a range of web-based services through hyperlinks. Popular commercial portals are provided by Yahoo! and MSN. Many portals can now be personalised to include personal favourites, bookmarks and feeds (see the description of netvibes in the 'aggregator' definition above). These portals can then be used as your homepage.

Profile: Information about a user of a social networking site. The profile is authored by the user, and usually the site will provide a template or prompt questions. These prompts often ask for user name, gender, location and favourite activities. Often the default on such sites is to make all details public, but as users are becoming more aware of security issues, such as identity theft, they may choose privacy filters so that they can control who has access to their details. For example, TroisTetes (TT) ironically describes himself on his Flickr profile as 'Witty, intelligent, good looking, modest and with a decidedly variable grip on reality, Che himself wears a TT T-shirt (true)'. He provides relatively little personal information apart from his email address and geographic location. As with many social networking sites, identity presentation is done far more through the content he uploads, than the profile he writes.

Social Networking Site: A website that sets out to support the development of a virtual community and provides opportunities for individuals to interact. This interaction may be loosely framed (as in Facebook) or focused on a particular 'social object' (such as videos in YouTube). Often social networking sites encourage users to create their own profile with biographical data, pictures, likes, dislikes and so on. Communication is also encouraged through comments, instant messaging and blogs.

Synchronous Communication: Communication that happens in 'real time'. Messages are responded to immediately and, although conversations may be 'saved' as files, it is not possible to edit them later (see Instant Messaging).

Tag: A label that can be given to a piece of information or digital object. Tags are devised by users and added to information they (or others) have uploaded. For example, YouTube users may add a tag to a film they have uploaded in order to help others understand what the film contains or even how they want the film to be understood—for example, 'comedy', 'spoof', 'fiction', 'American', 'advert'. Such tags will be added to the site's search engine and will assist other users in locating the film. (See also Tag cloud and Folksonomy).

Tag cloud: The aggregation of tags that have been assigned to the content of a website. This information is usually represented in a visual form, with the more popular tags shown in a larger font. The collection of tags will also reflect the number of times that a specific tag has been used on that site and thus reflect the most prevalent themes on that site. In this way, users can display something about the shared interests, even values, of those who use the site. (See also Tag, Folksonomy). Guy's tag cloud consists of words such as home, local, Sheffield, family, graffiti and so on, whereas Julia's tags include Sheffield, flowers, summer, streetart and Belrln. Each of their tag clouds reflects the image collection as well as the way they choose to describe them.

VLE (Virtual Learning Environment): An online space that has been specifically developed for learning (and teaching). Usually private (but not always), such spaces typically provide facilities for online discussions, collaboration over tasks and the embedding of 'learning objects' to aid discussion and learning. More complex sites contain blogs, wikis, video, podcasting (see Learning Platform). For example, the MA in New Literacies at the University of Sheffield is a customised learning space that allows asynchronous teaching to take place using Web CT. This means that learners can be distributed across a range of locations and time zones but still take part in the same 'learning activities'.

Widget: A generic term used to describe plug-ins or mini applications that can be incorporated in a web page. These are often a mixture of feeds and links. Examples are the Flickr 'badge' that shows an updated selection of photographs on your homepage, the Google search box and Last.fm's 'recently played' widget. Widgets are an example of interoperability at work.

Wiki: A collaborative web space that is based on the collective work of many authors. A wiki allows members to edit, delete or modify content that has been placed on the site. In this way, wikis grow as users add and refine the content to produce a shared resource. A popular example is www.wikipedia.org, whereas a more specific and relevant wiki is www.newlits.org.

For explanations of additional terms and concepts not listed here, we recommend searching www.wikipedia.org and www.howstuffworks.com.

REFERENCES

Alderman. N. (2008) 'We love doing the iPod shuffle', *Guardian* Newspaper 19 August 2008. Available at: http://www.guardian.co.uk/technology/2008/aug/19/randomness.shuffle. Retrieved 21 August 2008.

Alexander, R. J. (2006) (3rd edition) *Towards dialogic thinking: Rethinking classroom talk*. York: Dialogos.

Alfie (2005) 'London Underground bombing, trapped'. In *Moblog*. Available at: http://moblog. net/view/77571/. Retrieved 25 August 2008.

Bailey, F., and Moar, M. (2001) 'The Vertext project: Children creating and populating 3D virtual worlds'. In *Journal of Art and Design* (JADE) **20**:1, pp. 20–29 NSEAD.

Bakhtin, M. M. (1998) *The dialogic imagination*, M. Holquist (ed.). Austin: University of Texas Press.

Barnes, D. (1992) *From communication to curriculum*. Harmondsworth: Penguin.

Barton, D. (1994) *Literacy: An introduction to the ecology of written language*. Oxford: Blackwell.

Barton, D. (2001) 'Directions for literacy research: Analysing language and social practices in a textually mediated world'. In *Language and Education* **15**:2/3, pp. 92–104.

Barton, D., and Hamilton, M. (1998) *Local literacies: Reading and writing in one community*. London: Routledge.

Bauman, Z. (2003) *Liquid love: On the frailty of human bonds*. Cambridge: Polity.

Becta (2008) 'File sharing', May 2008. Available at: http://schools.becta.org.uk/index. php?section=is&catcode=ss_to_es_tl_uor_03&rid=12031. Retrieved 20 August 2008.

Benkler, Y. (2006) *The wealth of networks: How social production transforms markets and freedom*. New Haven: Yale University Press.

The Bentley Bros (2008) Bentley Bros Productions. On YouTube. Available at: www.youtube.com/bentleybros. Retrieved 27 August 2008.

Berners-Lee, T. (2006) 'Podcast and transcript of: *Developer works interviews: Tim Berners-Lee.*' Available at: http://www.ibm.com/developerworks/podcast/dwi/cm-int082206txt.html. Retrieved 20 August 2008.

Bestebreurtje, T. (2007) 'Second Life, a model for applications—generic web support for serious games in Second Life and beyond'. Amsterdam: VU University. Available at: www.cs.vu.nl/~eliens/multimedia/@archive/student/essay/**tom**-sl.pdf. Retrieved 20 August 2008.

Bigum, C. (1998) 'Solutions in search of educational problems: Speaking for computers in schools'. In *Educational Policy* 12:5, pp. 586–601.

Bigum, C. (2000) 'Managing new relationships: Design sensibilities, the new information and communication technologies and schools', paper given to the APAPDC Online Conference. Available at: http://www.apapdc.edu.au?2002/archive/ASPA/2000/papers/art 429.htm. Retrieved 22 October 2007.

Boud, D. (2001) 'Using journal writing to enhance reflective practice'. In L. M. English and M. A. Gillen (eds.), *Promoting journal writing in adult education: New directions in adult and continuing education.* San Francisco, CA Jossey-Bass, pp. 9–18.

Boutelle, J. (2005) 'Hey DJ—Web 2.0 and Remix Culture'. Available at: http://www.jonathanboutelle.com/mt/archives/2005/08/hey_dj_a_web_20.html. Retrieved 18 August 2008.

boyd, danah (2007) 'Why youth (heart) social networking sites: The role of networked publics in teenage social life'. In D. Buckingham (ed.), *MacArthur Foundation Series on Digital Learning—Youth, Identity, and Digital Media Volume.* Cambridge, MA: MIT Press.

Bryant, L. (2007) 'Emerging trends in social software for education'. In *Emerging technologies for learning* (Volume 2), pp. 9–18. Coventry: Becta . Available at: http://partners.becta.org.uk/index.php?section=rh&&catcode=&rid=13904. Retrieved 31 August 2008.

Buckingham, D. (2003) *Media education: Literacy, learning and contemporary culture.* Cambridge: Polity Press.

Burn, A., and Parker, D. (2003) *Analysing media texts.* London: Continuum.

Burnett, C. (2008) *Primary student-teachers' perceptions of the role of digital literacy in their lives.* Unpublished Ed.D. Thesis: Sheffield Hallam University, UK.

Burnett, C., Dickinson, P., Merchant, G., and Myers, J. (2006) 'Digital connections: Transforming primary literacy'. In *Cambridge Journal of Education* 36:1, pp. 11–29.

Cagle, K. (2006) 'A Web 2.0 checklist'. Available at: http://www.oreillynet.com/xml/blog/2006/02/a_web_20_checklist.html. Retrieved 22 August 2008.

Carr, D., Buckingham, D., Burn, A., and Schott, G. (2006) *Computer games: Text, narrative and play.* Cambridge: Polity Press.

Carr, W. (2000) 'Partisanship in educational research'. In *Oxford Review of Education* 26:3/4, pp. 437–449.

Carrington, V. (2008) '"I'm Dylan and I'm not going to say my last name"': Some thoughts on childhood, text and new technologies'. In *British Educational Research Journal* 34:2, pp. 151–166.

Carroll, S. (2008) 'The practical politics of step-stealing and textual poaching YouTube, audio-visual media and contemporary swing dancers online'. In *Convergence: The International Journal of Research into New Media Technologies* 14:2, pp. 183–204.

Cashmore, P. (2006) 'Second Life + Web 2.0 = Virtual world mash-ups' *Mashable Blog,* 30 May 2006. Available at: http://mashable.com/2006/05/30/second-life-web-20-virtual-world-mash-ups/. Retrieved 20 August 2008.

The Catsters (2008) 'The Catsters Channel'. On *YouTube*. Available at: http://www.youtube.com/user/TheCatsters. Retrieved 28 August 2008.

Charlieissocoollike (2007a) 'How to be English'. A video on *YouTube*. Available at: http://www.youtube.com/watch?v=BpWqCzru5zk. Retrieved 31 August 2008.

Charlieissocoollike (2007b) 'Winegum experiment'. A video on *YouTube*. Available at: http://www.youtube.com/watch?v=0rj2UJP3DRQ&feature=user. Retrieved 31 August 2008.

Clydehouse (2007) 'Comment on "Removed until further notice"' (Dr Joolz, 2007). Image on Flickr. Available at: http://www.flickr.com/photos/drjoolz/475484456/. Retrieved 30 August 2008.

Cocker, G. (2007) 'Jim Purbrick: Second Life and user-generated content'. *Gamespot UK*. Available at: http://uk.gamespot.com/news/6175694.html. Retrieved 20 August 2008.

Conservapedia (2008a) *Conservapedia: The trustworthy encyclopedia*. Available at http://www.conservapedia.com/Main_Page. Retrieved 28 August 2008.

Conservapedia (2008b) 'Wikipedi'. In *Conservapedia*. Available at: http://www.conservapedia.com/Wikipedia. Retrieved 28 August 2008.

Cope, B., and Kalantzis, H. (eds.) (2000) *Multiliteracies: Designs of social futures*. London, Routledge.

Daily Mail (2005) 'Emailing and texting fog your brain like cannabis'. *Daily Mail*, 22 April 2005.

Davies, J. (2004) 'Negotiating femininities on-Line'. In *Gender and Education* 16:1, pp. 35–49.

Davies, J. (2006) 'Escaping to the borderlands: An exploration of the internet as a cultural space for teenaged Wiccan girls'. In K. Pahl and J. Rowsell (eds.), *Travel notes from the New Literacy Studies: Instances of practice. Multilingual matters*, pp. 57–71.

Davies, J. (2007) 'Display, identity and the everyday: Self-presentation through digital image sharing'. In *Discourse, Studies in the Cultural Politics of Education* 28:4, pp. 549–564.

Davies, J. (2008) 'Pay and display: The digital literacy practices of online shoppers'. In M. Knobel and C. Lankshear (eds.), *Digital literacies: Concepts, policies and practices*. New York: Peter Lang, pp. 227–248.

Davies, J. (2009, forthcoming) 'A space for play: Crossing boundaries and learning online'. In Carrington and Robinson (eds.), *Contentious technologies: Digital literacies, social learning and classroom practices*. London: Sage.

Davies, J., and Merchant, G. (2007) 'Looking from the inside out—academic blogging as new literacy'. In M. Knobel and C. Lankshear (eds.), *The new literacies sampler*. New York: Peter Lang, pp. 167–197.

Davies, J., and Pahl, K. (2007) 'Blending voices, blending learning: Lessons in pedagogy from a post-16 classroom'. In J. Marsh and E. Bearne (eds.), *Literacy and social inclusion: Closing the gap*. Stoke on Trent: Trentham, pp. 118–138.

Dawkins, R. (2006) *The selfish gene. 30th anniversary edition*. Oxford: Oxford University Press.

de Certeau, M. (1984) *The practice of everyday life*. Berkeley, CA: University of California Press.

Dede, C., Clarke, J., Ketelhut, D., Nelson, B., and Bowman, C. (2006) 'Fostering motivation, learning and transfer in multi-user virtual environments'. *Paper given at the 2006 AERA conference: San Francisco*.

Donath, J., and boyd, D. (2004) 'Public displays of connection'. In *BT Technology Journal* 22:4, pp. 71–82. October 2004. Available at: http://smg.media.mit.edu/papers/Donath/PublicDisplays.pdf. Retrieved 30 August 2008.

Dovey, J., and Kennedy, H. W. (2006) *Game cultures: Computer games as new media*. Maidenhead: Open University Press.

Dowdall, C. (2007) 'Dissonance between the digitally created words of school and home'. In *Literacy* **40**:3, pp. 153–163.

Dr Joolz (2007) 'Removed until further notice'. Image on Flickr at: http://www.flickr.com/photos/drjoolz/475484456/. Retrieved 30 August 2008.

Duffy, P., and Bruns, A. (2007) 'The use of blogs, wikis and RSS in education: A conversation of possibilities'. Available at: http://eprints.qut.edu.au/archive/00005398. Retrieved 4 January 2008.

Dyson, A. H. (2003) *The brothers and sister learn to write*. New York: Teachers College Press.

Efimova, L., and Fielder, S. (2004) 'Learning webs: Learning in weblog networks'. Available at: http://blog.mathemagenic.com/2003/11/20.html#a844. Retrieved 4 January 2008.

Engestrom, J. (2007) *Microblogging: Tiny social objects. On the future of participatory media.* Available at: http://www.slideshare.net/jyri/microblogging-tiny-social-objects-on-the-future-of-participatory-media. Retrieved 31 May 2008.

Evangelista, B. (2003) 'Free music lessons thanks to file sharing'. *San Francisco Chronicle.* 11 November 2003. Available at: http://akamai.www.berkleemusic.com/assets/display/1111245/sanfran.pdf. Retrieved 19 August 2008.

Flickrblog (2007) 'Holy moly!' Post on *Flickrblog.* Available at: http://blog.flickr.net/en/2007/11/13/holy-moly/. Retrieved 31 August 2008.

Fors, A. C., and Jakobson, M. (2002) 'Beyond use and design: The dialectics of being in a virtual world'. In *Digital Creativity* **13**:1, pp. 39–52.

Frost, B. (2007) 'Rearchitecting the music business: Mitigating music piracy by cutting out the record companies'. In *First Monday* **12**:8, pp. 1–14.

Gee, J. P. (2004a) *What videogames have to teach us about learning and literacy.* New York: Palgrave Macmillan.

Gee, J. P. (2004b) *Situated language and learning: A critique of traditional schooling.* London: Routledge.

Gee, J. P. (2007) 'Pleasure, learning, video games, and life: The projective stance'. In Knobel, M. (ed.), *The new literacies sampler.* New York: Peter Lang, pp. 94–113.

Giddens, A. (1991) *Modernity and self-identity: Self and society in the late modern age.* Cambridge: Polity Press.

Gillmor. D. (2004) We the media: Grassroots journalism by the people for the people. Sebastapol: O'Reilly Media. Available at: http://oreilly.com/catalog/wethemedia/chapter/ch01.pdf. Retrieved 31 August 2008.

Greenbank, P. (2003) 'The role of values in educational research: The case for reflexivity'. In *The British Educational Research Journal* **29**:6, pp. 791–801.

Harris, R. (2000) *Rethinking writing.* London: Continuum.

Holloway, S., and Valentine, G. (2002) *Cyberkids: Children in the information age.* London: Routledge Falmer.

Ingram, A. L., Hathorn, L. G., and Evans, A. (2000) 'Beyond chat on the internet'. In *Computers and Education* **35**, pp. 21–35.

Jenkins, H. (2006a) *Fans, bloggers and gamers: Exploring participatory culture.* New York: New York University Press.

Jenkins, H. (2006b) *Convergence culture: Where old and new media collide.* New York: New York University Press.

Jenkins, H. (with Purushota, R., Clinton, K., Weigel, M., and Robinson, A.) (2006c) *Confronting the challenges of participatory culture: Media education for the 21st century.* Chicago: MacArthur Foundation. Available at: http://www.digitallearning.macfound.org/atf/cf/%7B7E45C7E0-A3E0-4B89-AC9C-E807E1B0AE4E%7D/JENKINS_WHITE_PAPER.PDF. Retrieved 27 August 2008.

Kent Local Authority (2006) 'e-Safety in Kent schools'. Available at: www.clusterweb.org.uk/docs/e-SafetyRecommendations_21.06.06.doc. Retrieved 31 August 2008.

Kerewalla, L., and Crook, C. (2002) 'Children's computer use at home and at school: Context and continuity'. *British Educational Research Journal* **28**:6, pp. 751–771.

Kinsella, S., Harth, A., and Breslin, J. G. (2007) 'Network analysis of semantic connections in heterogeneous social spaces'. Proceedings of the UK Social Network Conference, London, UK.

Kress, G. (2003) *Literacy in the new media age*. London: Routledge.

Kress, G., and van Leeuwen, T. (1996) *Reading Images: The grammar of visual design*. London: Routledge.

Kress, G., and van Leeuwen, T. (2000) *Multimodal discourses: The modes and media of contemporary communication*. London: Arnold.

Lamb, B. (2004) Wide open spaces: Wikis, ready or not. Available at: http://connect.educause.edu/Library/EDUCAUSE+Review/WideOpenSpacesWikisReadyo/40498?time=1219924835. Retrieved 25 August 2008.

Lankshear, C., and Knobel, M. (2006a) *New Literacies: Everyday practices and classroom learning* (2nd edition). Maidenhead: Open University Press.

Lankshear, C., and Knobel, M. (2006b) 'Weblog worlds and constructions of effective and powerful writing: Cross with care, and only where signs permit'. In K. Pahl and J. Rowsell (eds.), *Travel notes from the New Literacy Studies: Instances of practice*. Clevedon: Multilingual Matters, pp. 72–92.

Leander, K., and Mckim, K. (2003) Tracing the everyday 'sittings' of adolescents on the internet: A strategic adaptation of ethnography across online and offline spaces. *Education, Communication and Information* **3**:2, pp. 211–240.

Lessig, L. (2001) *The future of ideas: The fate of the commons in a connected world*. New York: Random House.

Lessig, L. (2004) *Free culture: How big media uses technology and the law to lock down culture and control creativity*. New York: Penguin.

Levy, P. (1997) *Collective intelligence: Mankind's emerging world in cyberspace*. Cambridge, MA: Perseus.

Lim, C. P., Nonis, D., and Hedberg, J. (2006) 'Gaming in a 3D multiuser environment: Engaging students in Science lessons'. In *British Journal of Educational Technology* **37**:2, pp. 211–231.

Linden Research (2008) 'Economic statistics (19 August 2008)'. Available at: http://secondlife.com/whatis/economy_stats.php. Retrieved 21 August 2008.

Livingstone, S., and Bober, M. (2004) 'Taking up online opportunities? Children's uses of the internet for education, communication and participation'. In *E-learning* **1**:3, pp. 395–419.

Luke, A., and Luke, C. (2001) 'Adolescence lost/ childhood regained: On early intervention and the emergence of the techno-subject'. *Journal of Early Childhood Literacy* **1**:1, pp. 91–120.

Markham, A. (1999) *Life online*. New York and Oxford: Almira Press.

Marsh, J. (2008) 'Out-of-school play in online virtual worlds and the implications for literacy learning'. Paper presented at the Centre for Studies in Literacy, Policy and Learning Cultures, University of South Australia: July 2008.

Marsh, J., and Millard, E. (2000) *Literacy and popular culture*. London: SAGE.

McCormick, R., and Murphy, P. (2008) 'Curriculum: The Case for a focus on learning'. In P. Murphy and K. Hall (eds.), *Learning and practice: Agency and identities*. London: Sage, pp. 3–18.

Mercer, N. (2000) *Words and minds: How we Use language to think together*. London: Routledge.

Merchant, G. (2001) 'Teenagers in cyberspace: Language use and language change in internet chatrooms'. In *Journal of Research in Reading* **24**:3, pp. 293–306.

Merchant, G. (2004) 'Imagine all that stuff really happening: Identity and children's digital writing'. In *Journal of E-Learning* **1**:3, pp. 336–340.

Merchant, G. (2007) 'Mind the gap(s): Discourses and discontinuity in digital literacies'. In *Journal of E-Learning* **4**:3, pp. 241–255.

Merchant, G. (2008) 'Writing the future in the digital age'. In *Literacy* **41**:3, pp. 118 –128.

Merchant, G. (2009) 'Literacy in a virtual world'. In *Journal of Research in Reading* **32**:2, pp. 38-56

Mortensen, T., and Walker, J. (2002) 'Blogging thoughts: Personal publication as online research tool'. In A. Morrison (ed.), *Researching ICTs in Context*. Oslo: Intermedia Report, pp. 249–279.

Murphy, P., and Hall, K. (2008) *Learning and practice: Agency and identities*. London: Sage.

O'Reilly, T. (2005) 'What is Web 2.0?' On *O'Reilly blog*. Available at: http://www.oreillynet.com/pub/a/oreilly/tim/news/2005/09/30/what-is-web-20.html. Retrieved 31 August 2008.

The Perklets (2008) 'The Perklets Channel. On YouTube. Available at: http://www.youtube.com/user/Perklets. Retrieved 28 August 2008.

Popular Music in Education (2008) *Popular music in education*. A Blog. Available at: http://blog.lib.umn.edu/chan0496/education/. Retrieved 19 August 2008.

Rheingold, H. (2002) *Smart mobs: The next social revolution*. Cambridge, MA: Perseus.

Richardson, W. (2006) *Blogs, wikis, podcasts and other powerful webtools for classrooms*. Thousand Oaks, CA: Corwin Press.

Rogoff, B. (1995) 'Observing sociocultural activity on three planes: Participatory appropriation, guided participation and apprenticeship'. In J. V. Wertsch, P. del Rio, and A. Averez (eds.), *Sociocultural studies of mind*. Cambridge: Cambridge University Press, pp. 139–164.

Schome (2007) *The Schome-NAGTY teen Second Life pilot: Final report*, Open University, May 2007. Available at: http://kn.open.ac.uk/public/document.cfm?docid=9851. Retrieved 20 August 2008.

Schroeder, R. (2002) 'Social interaction in virtual environments: Key issues, common themes, and a framework for research'. In R. Schroeder (ed.), *The social life of Avatars: Presence and Interaction in shared virtual environments*. London: Springer, pp. 1–19.

Selwyn, N. (2004) 'Reconsidering political and popular understandings of the digital divide'. In *New Media and Society* **6**:3, pp. 341–362.

Shirky, C. (2008) *Here comes everybody: The power of organizing without organizations*. London: Allen Lane.

Silverstone, R. (1999) 'Rhetoric, play performance: Revisiting a study of the making of a BBC documentary.' In G. Jostein (ed.), *Television and common knowledge*, pp. 71–90. London: Routledge.

Squire, K. (2002) 'Cultural framing of computer/video games'. In *Game Studies*. Available at: http://gamestudies.org/0102/squire/. Retrieved 12 May 2007.

Steinkuehler, C. (2007) 'Massively multiplayer online gaming as a constellation of literacy practices'. In *E-learning* **4**:3, pp. 297–318.

Steinkuehler, C. (2008) 'Cognition and literacy in massively multiplayer online games'. In J. Coiro, M. Knobel, C. Lankshear, and D. Leu, *New Literacies Research Handbook*. Mawah, NJ: Lawrence Erlbaum, pp. 611–634.

Street, B. (ed.) (1993) *Cross-cultural approaches to literacy*. Cambridge: Cambridge University Press.

Sunden, J. (2003) *Material virtualities: Approaching online textual embodiment.* New York: Peter Lang.

Tagg, K. (2005) 'Image on *Wikipedia*: "Trapped underground.jpg".' Available at http://en.wikipedia.org/wiki/Image:Trapped_underground.jpg. First uploaded by Alfie on *Moblog UK*, 7 July 2005 at http://moblog.net/view/77571/. Retrieved 25 August 2008.

Tapscott, D. (1998) *Growing up digital: The rise of the net generation.* New York: McGraw-Hill.

Technorati. (2006) *State of the blogosphere, April 2006. Part 1: On blogosphere growth.* Available at: http://technorati.com/weblog/2006/04/96.html. Retrieved 16 October 2006.

Thomas, A. (2007) *Youth online: Identity and literacy in the digital age.* New York: Peter Lang.

Thomas, S. (2004) *Hello world: Travels in virtuality.* York: Raw Nerve Books.

Trier, J. (2007a) ' "Cool" engagements with YouTube: Part 1'. In *Journal of Adolescent & Adult Literacy* **50**:5, pp. 408–413.

Trier, J. (2007b) ' "Cool" engagements with YouTube: Part 2'. In *Journal of Adolescent & Adult Literacy* **50**:5, pp. 598–604.

Turkle, S. (1995) *Life on the screen: Identity in the age of the internet.* New York: Simon Schuster.

Uncyclopedia (2008) 'Parody website: *Uncyclopedia: The content-free encyclopedia that anyone can edit'.* Available at: http://uncyclopedia.org/wiki/Main_Page. Retrieved 28 August 2008.

Walker, J. (2003) *Weblog definition.* Available at: http://jilltxt.net/archives/blog_theorising/final_version_of_weblog_definition.html. Retrieved 14 April 2008.

Wegerif, R. (2005) 'Reason and creativity in classroom dialogues'. *Language and Education* **19**:3, pp. 223–238.

Wellman, B. (2002) 'Little boxes, glocalization, and networked individualism'. In M. Tanabe, P. Besselaar and T. Ishida (eds.), *Digital cities II: Computational and sociological approaches.* Berlin: Springer, pp. 10–25.

Wellman, B., and Hogan, B. (2004) 'The internet in everyday life'. In W. S. Bainbridge (ed.), *Berkshire encyclopedia of human-computer interaction.* Great Barrington: Berkshire Publishing, pp. 389–397.

Wellman, B., and Hogan, B., Berg, K., Boase, J., Carrasco, J-A., Cote, R., Kayahara, J., Kennedy, L., and Tran, P. (2005) 'Connected lives: The project'. In P. Purcell (ed.), *Networked Neighbourhoods.* Berlin: Springer, pp. 1–50.

Wenger, E. (1998) *Communities of practice: Learning, meaning and identity.* Cambridge: Cambridge University Press.

Wikipedia (2005) '7 July 2005 London bombings'. In *Wikipedia.* Available at: http://en.wikipedia.org/wiki/7_July_2005_London_bombings. Retrieved 25 August 2008.

Wikipedia (2008a) 'Wikipedia: About'. In *Wikipedia.* Available at: http://en.wikipedia.org/wiki/Wikipedia: About. Retrieved 25 August 2008.

Wikipedia (2008b) 'July 7th 2005 London bombings'. In *Wikipedia.* Available at: http://en.wikipedia.org/wiki/7_July_2005_London_bombings. Retrieved 25 August 2008.

Wikipedia (2008c) 'Wikipedia: Neutral point of view'. In *Wikipedia.* Available at: http://en.wikipedia.org/wiki/Wikipedia:Neutral_point_of_view. Retrieved 25 August 2008.

Wikipedia (2008d) 'Wikipedia: Vandalism, Available at: http://en.wikipedia.org/wiki/Wikipedia: Vandalism. Retrieved 25 August 2008.

Willett, R. (2008) 'Consumption, production and online identities: Amateur spoofs on YouTube'. In Willett, R., Robinson, M. and Marsh, J. (eds.), *Play, creativity and digital cultures.* London and New York, pp. 54–71.

Williams, D., Wilson, K., Richardson, A., Tuson, J., and Coles, L. (1998) 'Teachers' ICT skills and knowledge needs: Final Report to SOEID'. Available at: http://www.scotland.gov.uk/library/ict/append-exec.htm. Retrieved 13 August 2008.

Willis, P. (2000) *The ethnographic imagination.* Cambridge: Polity Press.

Wormbins (2008) 'Our Wiki'. In *Wormbins*. Available at: http://wormbins.edublogs.org/. Retrieved 25 August 2008.

YouTube (2008a) *YouTube company history*. Available at: http://www.youtube.com/t/about. Retrieved 1 September 2008.

Zuckerberg, M. (2007) 'Facebook founder says social networking sites in it for long haul', interview: *Guardian Unlimited*, 18 October, 2007. Available at: http://www.guardian.co.uk/technology/2007/oct/18/. Retrieved 18 December, 2007.

INDEX

Colin Lankshear, Michele Knobel,
& Michael Peters
*General Editor*s

New literacies and new knowledges are being invented "in the streets" as people from all walks of life wrestle with new technologies, shifting values, changing institutions, and new structures of personality and temperament emerging in a global informational age. These new literacies and ways of knowing remain absent from classrooms. Many education administrators, teachers, teacher educators, and academics seem largely unaware of them. Others actively oppose them. Yet, they increasingly shape the engagements and worlds of young people in societies like our own. The *New Literacies and Digital Epistemologies* series will explore this terrain with a view to informing educational theory and practice in constructively critical ways.

For further information about the series and submitting manuscripts, please contact:

Michele Knobel & Colin Lankshear
Montclair State University
Dept. of Education and Human Services
3173 University Hall
Montclair, NJ 07043
michele@coatepec.net

To order other books in this series, please contact our Customer Service Department at:

(800) 770-LANG (within the U.S.)
(212) 647-7706 (outside the U.S.)
(212) 647-7707 FAX

Or browse online by series at:

www.peterlang.com